The
Happy Atom
STORY

READ
A Fantasy Tale
LEARN
Basic Chemistry

BOOK 1

IRENE P. REISINGER
ILLUSTRATIONS BY SARA K. WHITE
GRAPHICS BY IRENE P. REISINGER

Archway Publishing books may be ordered through booksellers or by contacting:

Archway Publishing
1663 Liberty Drive
Bloomington, IN 47403
www.archwaypublishing.com
1 (888) 242-5904

ISBN: 978-1-4808-6540-2 (sc)
ISBN: 978-1-4808-6541-9 (hc)
ISBN: 978-1-4808-6542-6 (e)

Print information available on the last page.

Archway Publishing rev. date: 9/18/2018

ARCHWAY
PUBLISHING

The Happy Atom Story
A Four Book Series

Book 1 The Atom and The Periodic Table

Guy learns to use the Periodic Table to understand the Atom.

Book 2 Bohr Models, Chemical Families and The Happy Atoms

The Bohr models show Guy the structure of the Atom. He meets the elements in the Chemical Families. As the book ends, Sodium discovers the secret of forming compounds—becoming Happy Atoms.

Book 3 Compound Formation, Chemical Formulas and The Valence Method

As Guy learns how the elements in each of the chemical families form compounds, he gets to understand the meaning of chemical formulas. It is then, that he learns a short method of constructing chemical formulas, the Valence Method

Book 4 Polyatomic Ions, Chemical Reactions and Balancing Equations

Guy watches Polyatomic Ions created and form compounds. He learns about chemical reactions and writing chemical equations. After that, the elements in the equations teach Guy to balance equations. See saws are used to demonstrate the concept. The book concludes with a parade of all the *principles of chemistry* marching in review.

DEDICATION
TO
Sister Regina Mercedes, CSJ
My Chemistry Teacher
St. Brendan's High School
Brooklyn, New York
and
John Carlin Ph.D.
My Inorganic Chemistry Professor
Fordham University
New York City
Thank you for helping me understand
Chemistry.

TO
MY LOVING PARENTS
Irene and John Murray
Thank you for my happy childhood.

TO
My wonderful husband, **Fred** and my five loving children
Terry, Mary, Kathy, Freddy and John
With All My Love

The Introduction
Table of Contents

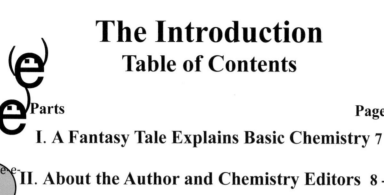

H_2O

Earth Metals

PO_4^-

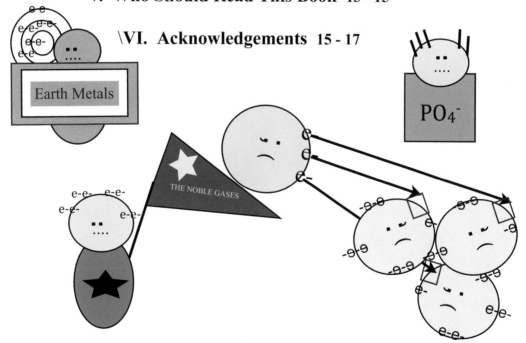

THE NOBLE GASES

Irene P. Reisinger

THE INTRODUCTION
A Fantasy Tale Explains Basic Chemistry

Would you believe you could read a fantasy tale and learn basic Chemistry? This is what will happen when you read *The Happy Atom Story.* In this book the principles of chemistry are woven into a fantasy tale. The narrative is about a young boy named Guy vacationing on the mountain for the summer with his family. He loves the mountain and nature all around him. Now he wants to learn all that chemistry can tell him about the unseen world beneath the visible world he loves. Guy's chemistry adventures begin when he meets Professor Terry whose magic Periodic Table has a hidden entrance to the fantasy world of Periodic Table Land. Guy's adventures take place in this mystical world. There, the elements, the silly electrons, the proper protons and all the little atoms eagerly share their knowledge of the world of chemistry with Guy. The reader learns chemistry along with Guy.

I used and refined this story teaching the basics of Chemistry to middle school students for a period of eighteen years. It was a success and my colleagues never gave up urging me to publish my Happy Atom story. Those in the Special Education Department who observed their students' success in learning chemistry were most persistent saying, "If you publish your story so many students will be helped." My middle school students loved the story and excitedly added their touches to it over the years. The success of *The Happy Atom Story* goes beyond the fact that the story holds the student's interest. Many concepts in chemistry depend on skills that students are missing. In this book these math skills are addressed where I've observed the need. Some are pure organizational skills that prevent errors. Read *The Happy Atom Story* and gain a better understanding of Basic Chemistry. Be ready to succeed in higher level chemistry.

The Happy Atom Story is published in a series of four books to keep the size of each book more manageable. Read the books in order as each builds on the previous book.

The Happy Atom Story
Book 1: The Atom and The Periodic Table
Book 2: Bohr Models, Chemical Families and the Secret of Compound Formation
Book 3: Compound Formation, Constructing Formulas and the Valence Method
Book 4: Chemical Reactions, Poly Atomic Ions and Balancing Equations

Irene P. Reisinger

II About the Author

The author was able to write this book because she has a thorough understanding of basic chemistry and the middle school child, and an imagination. Her education plus her life experiences prepared her to write this story. Her background in chemistry gave her the ability to turn the principles of chemistry into a story without distorting their meaning. She received the Gold Medal for highest honors in chemistry at graduation from Fordham University where she earned her undergraduate degree with a double major: Chemistry and Education. After graduation, she passed the New York City exam to become a licensed chemist. She worked as the chemist with a team of doctors doing research on the kidney for two years at New York University Medical Center in the Department of Renal Physiology. During this time she earned her Masters Degree in Science Education from New York University, Washington Square. Emphasis there was on creativity. Next she taught science for a short time in a junior high school in Brooklyn before marrying her college sweetheart who had now become an officer in the US Marine Corps.

The next part of her life contributed to her story-telling ability that was needed to create *The Happy Atom Story.* Her marine husband was gone six months to a year at a time giving her the opportunity to be creative raising her five little Reisingers. At one point she created a kindergarten in her home for Kathy, her third child, and nine other neighborhood five year olds. Then, while her husband worked at NSA and all her children were in school, she taught for five years in the Primary Department of Trinity Lower School in Howard County, Maryland. There she taught second graders how to write three paragraph compositions motivated by pictures. Her science background was utilized by having her enrich the science program for the Primary grades. All this definitely played a part in developing her imagination.

At that time her friends who were chemistry majors along with her in college were teaching in medical schools and universities. She wondered what she was doing involved with young children. Yet she loved what she was doing. Little did she know that she was being groomed to one day successfully teach chemistry to middle school students and eventually author this book. It is amazing how life experiences all blend together to be the perfect preparation for some future achievement as yet unknown. Her involvement with young children was preparing her to one day come up with the Happy Atom Story in that middle school classroom and to write this book..\

When her husband retired she began her career teaching middle school. She taught a course whose objective was to prepare middle school students to succeed in high school chemistry. She taught the middle school students basic chemistry for 19 years. It was in that 8th grade classroom that she discovered an extremely effective way to maintain the attention of middle school students, and at the same time have them learn chemistry. This was accomplished by weaving the principles of chemistry into a story. This story was used and refined over a period of 18 years and became the basis of this book, *The Happy Atom Story.* Without her life experiences and her knowledge of Chemistry this could never have happened.

Editing for Science Integrity

The background of the author was reported to assure the reader that the book contains valid science. The author's degree in chemistry provided her with far more knowledge than she needed to teach basic chemistry. For her undergraduate degree she had Inorganic, Organic and Physical Chemistry. Also, she had Qualitative and Quantitative Analysis which prepared her to be a chemist. But to assure you that the principles as explained in this book are accurate, she invited two chemists to read and critique her manuscript. It passed their inspection. The backgrounds of these chemists are described in the next section.

Joyce M. Donohue, Ph.D.—taught chemistry on every level high school, community college, university and graduate school. Presently, she's working for the EPA doing long term studies on water pollutants. In addition she is teaching chemistry in a local community college. Having taught high school Chemistry, Joyce's final comment was, "High school teachers will really appreciate the way your book prepares their students."

Mr. Robert L. Zimmerman, Jr., M.S.—is a chemist who was the Director of the Research Lab for Customs and Border Protection, a part of Homeland Security, until he retired. His new career, working for the American Association for Laboratory Accreditation to assess laboratories in different states to the *International Standard for Chemistry Labs.* After reading *The Happy Atom Story,* he liked especially the study suggestions the book provides.

I appreciated these professionals taking the time to read my manuscript assuring you that you are reading solid basic chemistry. It is important that a reader knows from the outset that the science they are reading is valid. This has been verified. Now to learn about the origin of Happy Atom Story.

Irene P. Reisinger

III. Origin of The Happy Atom Story

The Happy Atom Story was created one day in front of a class of 8th graders when the objective of the lesson was: Students will be able to explain how compounds are formed by either sharing or exchanging electrons. I stood in front of my fifth period class that met right after lunch. It was one hot, humid day in Virginia, and the air conditioning was not working. I had a gigantic Periodic Table hanging behind me on the blackboard. With pointer in hand I began my lesson. It was immediately obvious that I would have to come up with something dynamic if I was going to capture the students' attention. They were hot and sleepy after lunch.

The students had demonstrated that they had a command of all the chemistry needed to understand compound formation. I needed to figure out how to reach these hot and tired students. The rest of the unit hinged on learning today's lesson. My mind raced, searching for possibilities. I thought about the nature of the middle school child. They vacillate between childhood and adulthood. I decided that they needed a story—one that would help these kids learn how compounds were formed.

We had just finished learning about the chemical families. So, as I stood there looking at the Periodic Table, I suddenly saw it as the Land where the chemical families lived. I knew, the students would relate to these families and a story about the elements in these families.

I began saying, "OK, kids, I know it's hot and you're tired. Just sit back, relax, and listen. I'm going to tell you a story." A few students were nudged into a listening mode and I began.......

"Once upon a time there was a place called Periodic Table Land," and I moved the pointer to the words Periodic Table on the giant chart which was hanging on the blackboard. I paused for inspiration and went on with the story. "The chemical families, that we just learned about, live on these streets." I ran the pointer across the A Groups that would become the streets and across the B Groups which would become B Avenue. A few more heads rose from the desks. I then described the sad little elements with tears running down their dear little atom faces. Middle school students are very compassionate. So, they were interested in finding out why these atoms were so sad. At this point in the story, lots of heads began to rise off the desks. It was like watching a carefully choreographed ballet. Three heads rose in the middle of the class, two over there, several more up front. Soon, all heads were up, anxiously anticipating the next part of the story. I continued, "Well, I'm sorry to tell you that nothing made these dear little atoms happy. They were so sad that finally Sodium, the head of the Alkali Metal family, knew he had to search Periodic Table Land to discover how to make his family happy. As he searched, he found that elements everywhere were as sad as the elements in his family. He knew that if he ever found out how they could be happy, there would be a lot of happy atoms around.

Finally, Sodium reached 8th Street, the home of the Noble Gases. There they were all Happy Atoms. He stayed talking to the Noble Gases until he discovered their secret. He finally did. Their secret was, they all had complete outside energy levels. Sodium said

10

aloud, "If only all the atoms in Periodic Table Land could get complete outside energy levels, they all could be Happy Atoms too." On Sodium's way back to 1st Street, he figured out how this could happen........It was by exchanging or sharing electrons.

I drew a few sad atoms on the board and showed how these atoms could get the electrons they needed to become Happy Atoms. In no time, the story of compound formation was learned. The lesson's objective was achieved. Because the story was able to capture and maintain the students' attention, the students learned how compounds were formed. They were now ready to learn the rest of chemistry.

This was how the Happy Atom Story was born. Not only was the story a big hit with the students, but it also was effective in having them learn chemistry. Read the success stories that happened because of this story.

After I finished teaching compound formation, I continued the saga until all the principles of the unit were taught. It amazed me how one concept after another just fell into place as the story evolved to even explain how the Polyatomic Ions became compounds.

This story's success is what inspired me to write my book, *The Happy Atom Story.*

IV. The Story Proved Effective

Not only did the story save the lesson that hot, humid day in my 5th period class, it became the way I taught these units for the next 18 years. I had discovered an effective teaching tool to help middle school students learn chemistry. Students not only loved the story but they learned chemical principles that high school students find difficult to understand. Even my students with learning disabilities were doing well. My colleague, Barbara Brooks who worked with some of my students in the Special Education program was amazed that her students were able to construct chemical formulas because of the Happy Atom story. Barbara said, "In previous years, students used to dread going to science class when it was time to learn the unit on chemistry. This year, they can't wait to go to class to hear the next part of the Happy Atom Story, and they are truly learning Chemistry." She said often. "Irene, you have got to publish your Happy Atom Story. So many students will be helped if you do."

V. Success Stories with High School Students

Kati was another success story. Her mother had heard that chemistry was a really challenging subject, and she was concerned that her daughter would have problems learning the subject. She heard I was writing a book about chemistry, and asked me to help. Kati came and read my manuscript of *The Happy Atom Story* twice a week during the summer before she was to take high school chemistry. Helping Kati provided me an opportunity to settle the question: "Can a student read this book without teacher intervention and still understand it?" I knew that the Happy Atom Story helped students when I was there explaining it in the classroom as a teacher. I now needed to know if reading the book would be as useful a tool as the story was in the classroom. Kati read the book at my house each week; and after she read it, I tested her on the chemical principles she read. She proved that she thoroughly understood what I wrote without a teacher explaining it. Kati's contribution proved that the book can be read and understood without a teacher. Kati went on to achieve an A in high school Chemistry. She and her mother credited *The Happy Atom Story* for helping Kati succeed.

My granddaughter, Meredith Reisinger came to visit one weekend when she was taking high school chemistry. She was stressing about the thought of the test on compound formation that she was facing on the Monday right after her visit. I took her aside and explained a short version of the Happy Atom method of constructing chemical formulas. She aced the test.

IV. College Student Helped by Happy Atom Story

One of my former students told me how the Happy Atom Story played a significant role in her life, years after having learned the story in middle school. I was purchasing a Christmas present for one of my grandchildren in the gift shop of the Cracker Barrel Restaurant. The young girl who was wrapping the present said, "Aren't you Mrs. Reisinger, my 8th Grade Science teacher?" I said that I was. Of course, I didn't recognize her as she looked so much different now as a senior in college than she did in middle school. She continued. "I have to tell you how your Happy Atom Story influenced my life. I always dreamed of becoming a doctor. Then in freshman year college, I was taking chemistry, a requirement for my pre-med degree. I found it impossible to understand what my professor was trying to say. I had made the decision to drop chemistry, which meant I was giving up my dream of becoming a doctor. I went home giving serious consideration to my decision. I began to think how easy your Happy Atom Story made chemistry. After thinking about your story long enough, and how it explained chemistry, your story enabled me to figure out what my professor was trying to say. Because of your Happy Atom story, I was able to pass chemistry and remain in the pre-med program. In the Fall, I start medical school, and if it were not for your story, I would have given up my dream of becoming a doctor in freshman year. Thank you so much, Mrs. Reisinger."

VII. Decision to Write *The Happy Atom Story*

I was stunned. Never did I think that my Happy Atom Story would have effects beyond the middle school classroom. I knew the Happy Atom Story was a great way to teach chemistry to middle school students. I knew it would be great to share this method with other teachers. It definitely made it possible for middle school students to learn principles of chemistry usually considered difficult. I knew that the Happy Atoms made my *teaching* chemistry a successful and pleasant experience. It captured the students' attention and they learned. Before I heard that student's testimony about how the Happy Atom Story affected her life, I would have written my book, to help middle school science teachers teach chemistry successfully. Now because of this student's testimony I am writing the book with the hope of helping more students succeed in high school chemistry. If students are worried about taking chemistry because it has a reputation for being extremely difficult, I suggest they read my book before starting class. I still recommend the book to be read by middle school science teachers.

VIII. Who should buy this Book?

<u>Parents</u>: My neighbor is a parent who wants to see her child succeed. When I told her about the Happy Atom story she said, "Write that story in a hurry. My daughter is taking AP Chemistry in the fall. My husband and I do not remember enough chemistry to help." So she pointed out that this kind of a book would be eagerly received by parents who wish to help their child in such a challenging subject.

My granddaughter Kristine's husband, Phil has a Ph.D. in chemistry and works as a research chemist. He is interested in getting hold of my book and using it to interest his three little children in chemistry. He plans to read parts of it to them before they learn to read, letting them enjoy the pictures and the story. Later they can read it by themselves.

My daughter's neighbor wants the book to interest her children in chemistry. Both parents are physicians and one son is an avid reader. They want my book for this son to extend his interest in science to include chemistry.

My son's neighbor has a daughter who is going to take chemistry in the fall. The mom wants the book to use as summer reading to help her daughter feel good about taking a course in chemistry. She thinks the book will make chemistry seem less threatening.

<u>Home School Groups</u>: There are two home school communities who are interested in my book. They would like to use it at a young age to help their students become grounded in the basics of chemistry before they take high school chemistry. It can also help teachers get another perspective on chemistry, especially if they were not chemistry majors in college. I've spoken to many other parents who home school their children, and they see a great need for this kind of a book in their home school communities.

<u>Special Education Teachers</u>: There are requirements presently for special education teachers to demonstrate how they are differentiating their lessons to accommodate different learning styles. *The Happy Atom Story* definitely does this in the field of chemistry. The reader will appreciate how this book has a way of making theoretical subject matter visual and more understandable. Also stories help students understand concepts with explanations on their level, and stories help students remember the chemical principles longer. Students will love the story and become more successful. Therefore, the students' achievements will demonstrate that the teachers are indeed accommodating different learning styles.

Chemistry Teachers at Any Level: Chemistry teachers at any level would find value in reading *The Happy Atom Story*. It's a must read for Physical Science teachers in middle school. It affords the teacher a fresh outlook on how to explain so many principles in chemistry. Even high school teachers could learn alternate ways of explaining difficult principles through reading my book. I already wrote about how the Happy Atom Story helped my granddaughter, Meredith ace the test on constructing chemical formulas. She asked her teacher why she couldn't make it so simple the way her Grandmother explained it with the Happy Atom Story. Her teacher said, "I'd like to get a copy of her book." Well, at that time the book was not yet written. It is now.

Community college chemistry teachers would do well to read *The Happy Atom Story* to find alternate ways to explain chemical principles to their students. They might also suggest that their students having difficulty read this book. On this level there are many students who will not become chemistry majors who have much trouble understanding chemistry.

Students at Any Level: Students at any level would do well to read *The Happy Atom Story* during the summer before taking chemistry. This is true for students on all levels—middle school, high school, community college or even the university. The book provides a framework on which to hang the many more technical concepts provided at higher levels. After reading the book, students will enter the class with an overview of chemistry; and furthermore, they will not be struggling to see the relevance of the concepts being taught.They will understand where the new information fits in. A young friend suggested it should be on the Summer Reading List for students who might take chemistry in the Fall.

College Students with Chemistry as a Requirement: Many college students, who are not chemistry majors, are required to take chemistry for their degree. This is often a real challenge. Reading this book will prepare them to understand chemistry. By reading the book they will be establishing a base to build on.before they take the higher level chemistry course. A friend's daughter, whom I had helped with high school chemistry, went to the Air Force Academy. She told me that students who were English majors had to take chemistry because they were getting a BS Degree. They were having a really hard time. She said, "If they had read your book, it could have made their life much easier."

Students studying for a degree in Science Education would especially profit from reading *The Happy Atom Story*. This book would be a good resource if they someday have to teach chemistry as part of a science class.

The General Public: It is hoped by educators that the general public will increase their knowledge of all science so that they can have a better understanding of how science affects their lives. Research has shown that chemistry is the least understood of all the sciences. Reading *The Happy Atom Story* will give the reader an overview of what is basic to chemistry. I've spoken to many very intelligent people at different events and most of them verbalized that they had a hard time understanding what chemistry was all about when they took the class in high school. Others said, "I flat out failed that subject." One said, "I took the course and got a good grade, but I never did figure out what it was all about. I memorized everything." I've talked about the book with my doctors, and they agreed that chemistry was a difficult subject. They suggested that a book like mine is seriously needed. One podiatrist said, "I'd read it just for the heck of it."

Foreign Countries The internet indicates that foreign countries are looking for science books written in lay man's language. *The Happy Atom Story* fills this need in the area of basic chemistry.

IX. Acknowledgements

I'm indebted to the class of middle school students who were so hot and sleepy after lunch, that I was inspired to create *The Happy Atom Story.* I owe much also to the students in all my classes who encouraged me to add many details to the story making it grow in its appeal to middle school students.

I am indebted to all my fellow teachers who encouraged me to publish the Happy Atom Story. Word spread of how successful the story was, and so many teachers kept reminding me to publish the story. Mary Dvorznak, a music teacher and friend of Barbara Brooks kept after me, taking no excuses for delay in writing the book. She would say, "I don't want to hear any more excuses. Write that book." Annette Marsh, my close friend from day one of my teaching career at Graham Park, even called me when she moved to Texas to remind me to write *"that book."* She was such a dear friend that she never gave up asking when she was going to see the book published. She even sent me a complete course on discs explaining the publishing process titled, *How To Get Your Book Published.* I am so grateful for the many other teachers and friends, who heard about my story, and over the years kept asking, "Did you write that book yet?" Thanks to all of you for your encouragement.

My ability to create such an imaginative story in a middle school classroom can be traced back to the years I was involved in primary school education at Trinity Lower School in Howard County, Maryland, a school run by the Sisters of Notre Dame de Nemur. It was a wonderful time of my life, and to it I credit the development of my creativity. Thank you Sister Catherine Phelps for hiring me, nurturing me and being such a great principal. My creativity was encouraged also by Sister Shawn Marie Maguire, and my dear friend, Lynn Leaf. One cannot teach young children without becoming creative.

I have to mention in this section an appreciation for the fact that the people I needed to help me as I wrote this book just seemed to drop into my life. I'll just mention a few and there were many others.

I went to my chiropractor, Dr. Scott White. One day, just because in a split second I made the decision to tell him about the book, I had an illustrator. His daughter was a published illustrator. Sarah Kate White's illustration of Guy on the mountain and the night sky is loved by all. My daughter Terry loved it too, and also approved of Sara Kate's rendition of Professor Terry. I sincerely hoped she would, since Professor Terry was named after her. The way she drew Guy was exactly as I had imagined him.

I went to a bridal shower and the girl who sat next to me was a secretary to a copyright attorney, Christopher Collins. I asked him about copyright. When he saw what the book was about and that this was my first book, he took me under wing and guided me as a friend. He gave me the encouragement I needed to persevere.

I discussed the book with my Bridge group and found my editor, Cathy Turner. She was head of the English department at Woodbridge High School. She took an interest in my book and asked to edit my manuscript. I can never thank her enough for her endless hours of work.

My friends at church each contributed in different ways as I created this book. Claudia Brown was really special. When my illustrator was not able to continue, she volunteered to fill the void and drew many pictures for the book. She reviewed part of the book and made useful suggestions. She was so humble. I never knew she was an author as well as an artist. She was there when I needed her. Vickie Taylor spent a whole day with me editing Book 1, Part 1. Other close friends were there to listen to my challenges.

When I was having challenges using my computer, I went to the Apple Store and discovered mentors in a program called 1:1 who helped me learn the ins and outs of my Mac Book Pro laptop. When I began I knew nothing more than cut and paste. In that program I had three wonderful mentors: Samantha Kimbro, Christine Snopek and Randy Wood. They all had a hand in teaching me what I needed to know to write the book and create graphics to illustrate it. I remember each of them for something special that makes me say "Thank you for teaching me. Every time I click on an icon on the bar above, I thank Samantha who anticipated all that I needed to be handy as I wrote the book—the inspector, subscripts, superscripts, shapes, the color ball, text boxes, group and ungroup, font sizes, font types, the format bar, centering, bold/italics/underline and so much more. Randy was there the day I needed to make faces for the little elements I had created. He also taught me to use different shapes. This helped me create the flags and patches for the chemical families and those funny little polyatomic ions. I think of Randy when I look at my little elements and smile. Christine was there to show me solutions to all my frustrations. She taught me command Z, revert to, duplicate, export, open recent text boxes and best of all escape—to save me from all those connected boxes so annoying to eliminate. Later she was there to explain how to use my Apple to complete the submission requirements. She was so knowledgable.

Vaughn Treco was another person dropped into my life. At the Apple Store just when I needed him, he quickly observed that I needed a Periodic Table and proceeded to teach me. He had me create one box, fill it with the information about one element carefully creating the proper spacing. Then he had me duplicate it 118 times times, and join the boxes together, creating the table. The final and painstaking step was to replace the information in each box with the correct information for the element that belonged there. When finished I had a Periodic Table. Some of the Periodic Tables needed coloring. That was another lesson. I am very grateful for Vaughn. Recently, my original helpers were all promoted and moved on. Then, Alex Oldershaw was there to help. He was able to solve the problems that I faced. After a while I was able to use the skills I had acquired to solve difficult problems on my own. Everyone at Apple is anxious to see the book published because they were there to watch it develop. They always gave me the feeling that they truly enjoyed helping me. This made my visits to the Apple store very pleasant. They were wonderful ambassadors for the Apple Corporation. I have nothing but praise for Apple because of my mentors in 1:1. Apple really takes care of the people who buy their products.

In the area of editing I received help for a short time from my granddaughter Maureen Robbs and her husband Josh, both busy professional editors. When Maureen could no longer help, my daughter, Terry Halsema took over. She was my right hand person through the whole writing experience. Terry is a multi-talented person whose advice and judgement I value. She edited and re-edited every time I made changes. My daughter, Mary Pat Butrym, helped me edit my manuscript right before sending it to the publisher. My trip to Wisconsin to visit family earned me two important bits of knowledge. My daughter, Kathy Palmer, showed me how to get images moving with the text. Nick Stevens, a computer engineer, inspired me to make computer generated charts.

My son, Fred Reisinger, read the book and said, "Where was this book when I took high school chemistry?" He liked how I simplified complex principles.

Rita Slessinger introduced me to two authors, Tabbie Mann and Brenda Smith Warren who shared with me their experiences. I met them at the beginning of my venture, when encouragement was very needed and appreciated.

Finally, I was encouraged by Einstein's quote, that I heard on TV : *Logic will get you from A to Z. Imagination will get you everywhere.* Imagination helped me achieve the objective of my lesson on that hot day in a Virginia middle school classroom when I created the Happy Atom Story. It also changed the way I taught chemistry for the rest of my teaching career. Now imagination has helped me create this book.

Thanks to everyone that I neglected to mention. Just know that I gave serious consideration to any suggestion, even casual comments, that you made. You all had an impact on the final version of this book. Thank you, thank you, thank you!

Irene P. Reisinger

BOOK 1
PART 1
 *TABLE OF CONTENTS

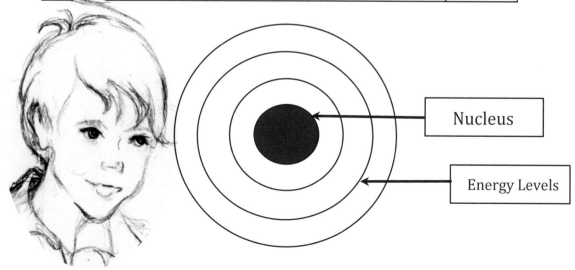

Nucleus

Energy Levels

Irene P. Reisinger

BOOK 1
PART 1
Guy's ChemistryAdventure Begins

It was Guy's family's first night in their summer cabin on the mountain. Guy peeked out the back door and saw that it was finally dark enough to see the stars. He couldn't wait to get to his favorite spot on the mountain overlooking the valley. Full of excitement and anticipation he walked quickly, but carefully, over the rocks and ruts in the tree-lined path. Finally, he spied his private rock which was inclined to just the right angle to comfortably scan the night sky. As he cast his eyes toward the twinkling stars above, he felt a gentle breeze brush across his face and the aroma of pine reminded him how much he loved his mountain. Staring at the night sky, his imagination ran wild thinking of the planets twirling around in space millions of miles away. He spotted a few of his favorite constellations. Seeing Orion and the Big Dipper Guy said out loud to himself, "Oh, how magnificent my world is! I just love sitting here looking at the stars in the night sky. It's all so awesome. " Guy felt so in tune with the entire universe.

Across the dark sky, he saw a shooting star. He remembered a rhyme his father told him when he was little. *Star light, star bright, first star I see tonight. I wish I may, I wish I might, have the wish I wish tonight.* He thought, it wasn't the first star I saw tonight, but it was the most exciting. My wish is to learn the secrets of my magical world.

Suddenly, a streak of light drew Guy's attention to the top of the pine tree near him. As the tree began to sway back and forth, Guy saw a little star poke his head through the branchs of the pine tree. In an instant the star jumped down and landed beside him with a thud.

"Hi Guy! I'm Wish Star," he said jumping up and down waving his magic wand. "Comet Star sent me here to make your wish come true. I can teach you a lot about your wonderful world. We'll begin tomorrow. Here's a book that Comet Star wanted you to have."

Guy saw how thick the book was, and he was concerned that it would take forever to finish it.

Wish Star noticed and said, "Don't worry it's a magic book. You will love how it works. You'll see. I'll show you tomorrow."

Guy tucked Comet Star's book into his backpack and got ready to leave. Filled with wonder and awe, Guy walked home beneath a blanket of twinkling stars. He told himself, "My own special Wish Star! A magic book! Wow! It looks like this is going to be a special summer." Guy had not the slightest idea how special it would be.

Early the next morning, Guy woke up and hiked to his second favorite spot on the mountain. It was a grassy field along side a cool stream lined with stones and vibrant wildflowers. He spread his blanket on the grass beneath a shady tree. As he rested his head against the tree, he listened for the sounds of the forest stirring as the sun rose in the sky. He heard a squirrel scurrying up a tree and the splash of a fish jumping out of the water only to dive deep again. The sun glistened off the stream. A glitter off one of the rocks caught his attention.

He walked towards the stream where he found three coins. He picked them up and turned them over in his fingers realizing that someone else must have been enjoying this wonderful spot. He rested his back on the blanket and looked up at the fluffy, white clouds that decorated an otherwise clear blue sky. "What a beautiful world this is!" he exclaimed. He was ready to find out more about his magnificent world.

Guy remembered the book that Wish Star had given him. It was a magic book, and it would help him understand the world he loved. So Guy pulled the book out of his backpack and tried to flip it open. No matter how he tried the book would not open.

This time when he tried to open the book Wish Star appeared out of nowhere and landed on Guy's blanket right next to him.

"Hi Guy! I told you I'd find you. The moment you tried to open the book, it sent out a signal to me, and here I am." Wish Star tapped the book with his magic wand and the book opened to a special page. He explained to Guy, "When I tap the book, it opens, and out floats the information we need. Then, whenever we finish a topic, the information slides back into the book, and the next chemical principle you need to learn jumps out. My magic wand is, to put it simply—magical! My magic wand can make the book give us the information we need."

Guy Learns About Matter

Wish Star said, "I guess you wonder why the book opened to this page." Then he answered his own question, "The book knows what you need to know. Up in the universe and also on earth, well everywhere, there are special words like matter, mass, and space whose exact meaning you need to know to understand everything else in science."

Guy paused to consider what Wish Star had said. Then Guy nodded to show he understood and said, "Sounds like the book will help me understand everything I need to know to begin learning about my wonderful world. So let the learning begin. I'm ready."

Wish Star touched the book with his magic wand, and out jumped a definition, and it floated in the air in front of Guy,

Matter is anything that occupies space and has mass.

Wish Star explained, "Scientists call the *stuff* in living and nonliving things **matter**. So scientists use the word matter instead of things and stuff."

Guy leaned back against the tree and looked at the landscape around him. After much thought, Guy told Wish Star, "I guess that means those squirrels, rocks, grass, and everything around me is matter." After pausing a few seconds Guy added, "I guess, even my body is matter!"

Mass or Weight

Wish Star, pleased that Guy understood **matter,** looked at Guy and directed him to the next idea. "Think of this. If you think you are too heavy or too light, you climb on the scale to find out your weight. You people here on earth usually describe how heavy things are as **weight.** Scientists call how heavy things are **mass.** You have to know the difference between weight and mass. Guy, how heavy we are depends on how much matter we have in our body. Matter is the stuff things are made of and how much stuff we have in our body is what makes us heavy or light. Scientists call this mass."

One tap on the book and out popped the definition of the word **mass:**

Mass is the *amount* of matter in an object.

Then out came the definition of the word **weight.**

Weight is the *relative mass* of matter caused by the pull of gravity.

Again it hung in space in front of Guy. Matter's definition had slid back into the book silently. Wish Star explains, "If we climb on a scale and weigh a lot, we are heavy. That means there is a lot of matter in our body. Matter is the 'stuff' that makes our body heavy—like bones, fat, muscle and everything else inside our body. If the scale says we weigh little, we don't have much matter in our body. **Scientists call this mass, not weight.**"

Guy said, "I got it. The matter in our body makes us heavy or light. Scientists call this our mass but we call it our weight." Guy turned to Wish Star and asked, "Why don't scientists call it weight?"

Wish Star decided to stand on his head as he explained   the answer. Guy laughed and said, "You're funny Wish Star."

Wish Star said, "To scientists mass and weight have totally different meanings. **Mass** describes the *amount* of matter in something. Mass is the same everywhere. **Weight** describes *the effect of the pull of Gravity* on matter. Weight changes where gravity is different. So when scientists call something weight, it's to let the reader know that gravity is involved."At that point gravity pulled Wish Star down, and Wish Star said, "I guess gravity wanted to remind me of his presence."

Guy said, "I'm not quite sure I understand. Could you give me an example."

Here's the story Wish Star told. "An astronaut climbs on the scale before the rocket takes off for the moon and says, "My weight is 180 pounds." He really means his mass is 180 pounds. That's how much matter he has in his body now on earth, and the same amount of matter he will have in his body when he lands on the moon. When the astronaut climbs on a scale on the moon, it will say his weight is 30 pounds because the pull of gravity on the moon is 1/6 of what it is on earth. His weight changed with gravity. Because pull of gravity on his body was less on the moon, his weight was less on the moon. His mass however remained unchanged. He still had the same bones, fat and muscle in his body. His mass was still 180 pounds. This is because he still had the same amount of matter in his body as he did on earth.

He continued, "Weight is a description of the effects of gravity pulling down on matter. Weight represents the force of gravity pulling on the mass of our body, the amount of matter in our body. Weight is different where gravity is different. Mass is the same everywhere. On earth we we use the terms weight and mass interchangeably because gravity is essentially the same everywhere on earth. Scientists only use the term weight where gravity is involved."

Guy responded, "I never knew that before, but now I understand."

Wish Star tapped the book with his magic wand. Mass and weight went back in. Then, out popped the word space and hung right in front of Guy.

Space

Wish Star tapped the book once more, and out floated the definition:

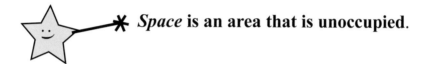

Space **is an area that is unoccupied.**

"Everything that is matter takes up space that once was unoccupied." Wish Star explained, "Imagine there is a bench that only has enough room for five kids. When another kid comes along who wants to sit down you say, 'Sorry, there's not enough space for you to sit here. However, we can squeeze together to create more unoccupied space for you'——**Two things cannot occupy the exact same space at the same time."**

Guy thought about this principle and said, "I guess I always knew this."

Wish Star summarized: "**Matter** (the stuff in things) has **mass** (how heavy that matter is) and matter takes up **space** (how much space depends on how large the matter is). It's simple, but you have to understand the words—**matter, mass, and space.** That's the language of scientists."

Wish Star asked Guy, "Do you have any questions about Matter? We have found out what it takes to be considered Matter."

Guy said, "I do have a question. I don't understand whether or not air is Matter."

Guy's Question: "Is Air Matter?"

Wish Star said, "Watch Guy. Comet Star's book is absolutely intriguing. It will answer your question with a flick of my magic wand." One flick of Wish Star's magic wand and the answer floated out of the book into the air space in front of Guy.

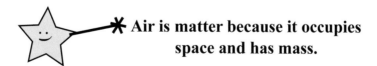

Air is matter because it occupies space and has mass.

Guy objected, "But you can't even see air."

The book clarified, "If you pump air into a balloon, the balloon becomes larger. So, obviously air takes up space. If you had a sensitive enough scale, and weighed the balloon before and after blowing the air into it, the balloon would weigh more with the air in it. So, air has mass. This proves air is matter. Even if you can't see it, air has mass, and it takes up space."

Guy took a deep breath of that refreshing mountain air and smiled. He felt good that he had learned three new words to describe all the matter around him. He turned to Wish Star and said, "Matter, mass, and space are new ways for me to think of my surroundings. I wonder if there is anything in my world that is not matter?"

Guy Learns About Energy and Forces

Suddenly, Guy heard the rumble of thunder in the distance. "Come on, Wish Star, we've got to get to the cabin. When you hear thunder, lightning is not far away. Being out here is dangerous." He gathered up his belongings. Wish Star climbed on Guy's shoulder, and they dashed off to the cabin. Quickly Guy turned on his parent's antique radio that confirmed a storm was approaching. He looked out the window. As he saw a flash of lightning Guy said, "We just got into the cabin in time." The wind had picked up in force. He saw a branch falling to the ground after it was knocked off the tree by the force of the wind. He said, "I guess nature is answering my question: Is anything in this world not matter?"

The wind jogged Guy's memory. He shared with Wish Star, "I remember a book my father read to me when I was little. It was all about forces and energy in our lives. It had so many beautiful pictures on every page. I remember that the purpose of the book was to show that there is something else in this world besides matter."

Wish Star knew quite a bit about this. So he shared what he knew. "Of course Guy, besides matter, there's energy, forces and motion. We have just experienced several of these. For instance, your parent's old radio here is matter; however, the radio waves it picks up and changes to the radio announcer's voice are energy. It's like your parent's smart phone, Guy. It picks up those same energy waves and changes them to the data they receive. Thunder is sound energy. Lightning is light energy. Then, there is the motion of the wind that knocked that branch off the tree. Then there's the force of gravity which pulled that branch to the ground."

Energy and Forces vs Matter

Wish Star said, "Let's look at what Comet Star's book has to say." One tap with the magic wand, and the words floated out of the book and hung in the air in front of Guy.

Energy and forces don't have mass or take up space.

Energy and forces are not matter because matter has mass and takes up space. Energy and forces do not have mass nor do they take up space.

Guy found the book his father once read to him. "Look Wish Star. here it shows an electric cord draped over a scale. You are asked to notice that the electric cord does not weigh more when more electricity is pumped into it. It also, asks you to observe that the wire does not become larger when more electricity is pumped into it the way a balloon gets larger when more air is pumped into it. Electricity does not have mass nor does it take up space. This experiment with the electric wire proves electricity is not matter because it does not take up space, and it does not have mass. Electricity is a form of energy. It is not matter."

Wish Star hopped onto the windowsill and had something to tell Guy. "I can tell you first hand that the stars who sent me to you have lots of energy—light energy and heat energy. Your sun is the only star close enough that you can feel its heat, but all stars give off heat as well as light. Notice they turned off my heat and toned down my light. If they did not do this, I couldn't help you."

Guy's Decision to Learn Chemistry

There is so much to learn about my world, thought Guy—energy, forces, matter—I want to learn it all. Wish Star bounced right over to Guy. He looked seriously into his eyes as he proposed a decision that Guy had to make. "You have to decide now whether you will study matter, or energy, motion, and forces. **Chemistry is the study of matter. Physics is the study of how energy, motion, and forces interact with matter**. Guy made his decision, "Physics sounds interesting, but I would rather learn chemistry first."

Wish Star said, "That's a good decision Guy. Since Physics is the study of the effect of energy, motion, and forces on matter, it makes sense to study chemistry first. That way you will have a firm understanding of matter before you begin Physics. I know a great person to help you learn chemistry. Follow me." Off they went to the university at the base of the mountain.

On the way down the mountain Guy looked at the trees and flowers along side the path. He thought, "It won't be much longer before I'll get to know more about you."

Guy Meets Professor Terry

When they entered the lab Professor Terry immediately recognized her old friend, Wish Star, and greeted him warmly. Wish Star introduced Guy to Professor Terry saying, "Well Professor, this is my little friend, Guy who is spending the summer with his family up on the mountain. He really wants to learn chemistry. Perhaps you would have time to help him. Comet Star sent me to help Guy, but I can't stay all summer."

Professor Terry looked down at Guy and smiled. "Well Guy, I'm so glad to meet someone as young as you who is curious about your world!" Guy beamed. He was so happy that Professor Terry was pleased to meet him. He hoped that she would agree to help him since Wish Star couldn't stay all summer.

Professor Terry said, "Are you sure you want to learn about chemistry?"

Guy responded, "I love the night sky, the planets, and nature on this mountain. What I am really hoping to learn are the secrets beneath the surface of the world I see."

Professor Terry said, "Well that's what chemistry is all about. All those things in the universe that you love are made up of atoms that are so small that you can not see them. Do you really want to learn chemistry Guy?"

Guy said, "Oh yes, I really am ready to learn as much as I can about my world. I want to learn what this world is made of."

Professor Terry was delighted to hear that Guy was serious about learning chemistry. So she looked thoughtful and said, "I have some free time right now, and I'd be happy to help you. Is there anything special you want to learn?"

Wish Star flew up onto the lab table and spoke up, "I've taught Guy about matter. I think learning about atoms is a great place to start, and Guy is ready."

Professor Terry agreed, "Good idea! Learning about the atom is just a perfect place to begin."

Guy Learns About the Atom

Professor Terry gladly responded, "Of course! Lets learn about the atom. **The atom is the basic unit of matter in the universe**. Come on over to the blackboard," Professor Terry began explaining. "The atom has two parts and there are particles in these parts." She picked up a piece of chalk and began to draw the parts of the atom first. The diagram—**The Parts of the Atom**. was soon drawn and labeled on the board. "Look at the diagram Guy. These are the parts of the atom–the nucleus and the energy levels.

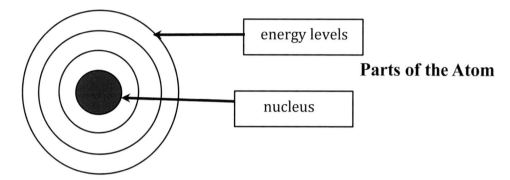

Parts of the Atom

Now look carefully." She pointed to the dark center of the atom, "This is the nucleus." Then pointing to the rings around the nucleus, she said, "These are the energy levels."

"Next I'm going to tell you about the particles in the atom—their location and the symbols we will be using for them. On the blackboard she drew another atom and labeled it—**Particles in the Atom**. The **electrons** are particles that can be found in the **energy levels.** There are two particles that are found in the **nucleus of the atom**. These particles

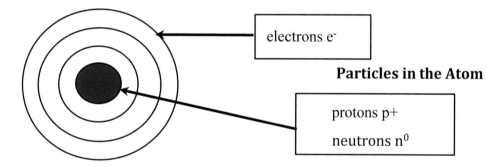

Particles in the Atom

are the **protons and** the **neutrons**."

Professor Terry added, "In studying the nucleus of the atom, scientists have encountered things like quarks, leptons and several others with equally charming names. One kind of quark is actually called a charm. None affect the mass of the atom. You will enjoy reading about these some day after you know the basics of chemistry. That's what I would be happy to teach you now—the basics of chemistry."

Guy was excited that Professor Terry was willing to teach him about chemistry.

Guy Learns How Atoms Are Different

"Guy, the diagrams of the atom that I drew for you are the just basic models. There are 118 different kinds of atoms that have been discovered and they are called elements. Each element has an atom that is different from the atom of every other element because it contains a different number of protons in its nucleus."

Guy Learns About Elements

"Each element has **a name, a symbol, and an Atomic Number**. Each element is listed on a chart called **The Periodic Table of the Elements**. On this chart, you can learn a great deal about each of the elements. I will give you a small chart with the first twenty elements that you can find on The Periodic Table. On this small chart you will find the element's name, symbol and Atomic Number." Professor Terry went over to the lab table, rummaged through a stack of papers, and finally found what she was looking for. She

Professor Terry's Element/Symbol Chart
Atomic Number/ Name of Element/ Symbol

1	Hydrogen	H	8	Oxygen	O	15	Phosphorus	P
2	Helium	He	9	Fluorine	F	16	Sulfur	S
3	Lithium	Li	10	Neon	Ne	17	Chlorine	Cl
4	Beryllium	Be	11	Sodium	Na	18	Argon	Ar
5	Boron	B	12	Magnesium	Mg	19	Potassium	K
6	Carbon	C	13	Aluminum	Al	20	Calcium	Ca
7	Nitrogen	N	14	Silicon	Si			

returned with the Element/Symbol Chart, giving it to Guy.

"Guy, I'm also giving you another paper which contains hints for easy ways to memorize the symbols of the elements. **Read the Helpful Hints paper first.** Then learn the symbols for these 20 elements, and it will give you a head start. Once you know these symbols, come back to visit me, and we will talk more."

Helpful Hints for Learning the Symbols

- First, memorize the elements whose symbols come from a foreign language.
Sodium–Na, Potassium–K
- Next, memorize the elements with single letter symbols.
Boron–B, Carbon–C, Fluorine–F, Hydrogen–H, Nitrogen–N, Sulfur-S, Phosphorus–P

- All the other symbols are the first 2 letters of the element's name
Aluminum–Al, Argon–Ar, Beryllium–Be, Calcium–Ca, Helium–He, Lithium–Li, Neon–Ne, Silicon–Si

EXCEPTIONS
Chlorine–Cl and Magnesium–Mg

Wish Star had been quiet for a while. Now he spoke up. "Guy, I think you should hear what Comet Star's book has to say about elements."

Comet Star's Book Tells Guy About Elements

He tapped the book with his magic wand and out floated the definition of an element. It hung in the air space right in front of Guy.

An Element is a pure chemical substance consisting of only one type of atom.

It hung in front of Guy to give him enough time to think about it. Then out floated a paragraph of information about elements.

Each element is distinguished by its **name**, its **symbol**, and its **Atomic Number**. The **Atomic Number** tells us how many protons are in that atom. That's what makes each element unique. Every element has a different number of protons in its atom. It also tells the number of electrons it should have. Out floated some examples.

Guy noted, "Professor Terry mentioned the same facts. They must be important."

Examples:
* Oxygen, symbol O, Atomic Number 8, has 8 protons and 8 electrons.
* Carbon, symbol C, Atomic Number 6, has.6 protons and 6 electrons.
* Chlorine, symbol Cl, Atomic Number 17, has 17 protons and 17 electrons.

More Facts About Elements

Wish Star tapped the book one more time. Out jumped more facts. Wish Star smiled.

* 118 different elements have been discovered, four of these just recently.
* Each of these elements have been given an Atomic Number and a symbol and placed in a chart called the Periodic Table of the Elements.
* Rules: Each element's symbol *begins with a capital letter*. If there is a second letter, it is ***always*** *lower case*.

After hiking back to his family's cabin, Guy read over the helpful hints. As his mother prepared lunch, Guy sat down by the window and began to memorize the symbols. Guy was happy to begin his study of chemistry.

****

After lunch, Guy took a walk to think about what he had learned. He was so excited to think about his world made up of tiny little atoms. He looked at the trees, the stream, the squirrels, the rocks and grassy hills and imagined them with all their tiny atoms invisible to his eyes. In his mind, he saw their tiny little electrons whirling around a nucleus containing protons and neutrons. He was filled with awe.

Gradually, his mind drifted to the magnificent variety in his world. It is amazing that things so entirely different could be made up of little atoms that are basically similar. They all contain a nucleus with protons and neutrons, and electrons that revolve around it. The question lingered in Guy's mind, "How can there be such a variety in this world? How can there be thousands of different kinds of things that look so entirely different when there are only 118 elements?"

The Amazing Variety In Our World

As Guy continued down the path, he heard someone approaching. Soon Professor Terry rounded the bend. She waved and said, "You left your keys in my lab, and I thought you might need them."

"Thank you, Professor," Guy said. "I didn't even notice they were missing because my parents were at home to let me in. Thank you for finding them. I am also happy to see you. I have a question."

"Of course," Professor Terry said, "But first, let's sit on those large rocks beside the stream."

"Great!" Guy responded. "I was looking around and noticing the tremendous variety in our world. I wondered how this was possible if everything in the universe is made up of little atoms that are not so different, and there are only 118 of them. How can there be thousands of things that look so different?"

Some Elements Created in the Lab

Professor Terry began, "Before we go any further, I have to tell you that there are even fewer elements that contribute to this diversity in our world. **After the element Atomic Number 92, which is Uranium, the rest of the elements are not found in nature but have been created in a laboratory.** Some of them stay in existence for only seconds. So they are not around long enough to contribute to the diversity of matter at all."

"Now let me begin to answer your question about how there can be so many different things in this world when there are so few elements. Think about the butter on your dinner table. Essentially, it's made up of only three elements: Carbon, Hydrogen and Oxygen. If you join those same three elements together in different ways, you get thousands of different forms of matter—matter that is decidedly different from each other. Here are examples of matter made up of the same three elements, Carbon, Hydrogen and Oxygen: sugar that sweetens your food, gasoline that powers your car, and rubbing alcohol used in hospitals. Some products are made up of only two of the elements in butter: Carbon and Hydrogen: propane gas that lets you barbecue hot dogs in your back yard, methane which is swamp gas, and butane that fires up a lighter. If you throw in just one more element Nitrogen you have the make up of many parts of the human body. The thousands of possible ways that elements can join together is part of the explanation of how there are so many different forms of matter from so few elements. This is one answer to your question. It's the story of chemistry."

Wish Star was on the sidelines listening. He couldn't be quiet any longer. "Comet Star's book has several more ways to explain how there can be so many more things in this world than the mere 92 elements can explain."

Professor Terry said, "I'm so glad you're here. I have another meeting to attend. So Wish Star take over from here. I'll be back in an hour."

Wish Star leaped from the other side of the bushes with a streak of light trailing behind him. He landed in front of Guy who was still sitting beside the stream. "Before I start teaching you more reasons things can look different, we have to review a few things."

"Let's do a little review. Matter is stuff that has mass and occupies space. Then you learned matter was made up of little atoms. Finally there are only 92 different kinds of atoms found naturally in our world. Now let's see what I can add to this. I'm going to show you another way matter is different." Waving his magic wand, he touched it to Star Comet's book and out popped words that hung right in front of Guy.

States of Matter
Solid, Liquid, and Gas

Guy was enchanted seeing these words come out of the book and float in front of him once again. He thought they were finished with Comet Star's book.

Wish Star said, "I know you have heard of solids, liquids and gases before in your every day life. Some things are solid, like these rocks we are sitting on. Solids have a definite shape. Other things are liquid, like this stream." Wish Star reached down with a cup, scooped up some water, and demonstrated how liquids can be poured. "Then too I'm sure you know that air is a gas. With a gas like air, you can take a stick or your hand and move it right through this form of matter, just because it's a gas. I'd like you to make a list of some other solids, liquids, and gases you know about."

Guy said, "I can think of a few." So he wrote:

* **SOLIDS:**   Rocks, trees, leaves, bones, teeth and baseballs.
* **LIQUIDS:** Water, soda, milk, vinegar, and gasoline.
* **GASES:**   Air, oxygen, propane, carbon dioxide, hydrogen and helium.

Wish star tapped the book again and out floated three boxes.

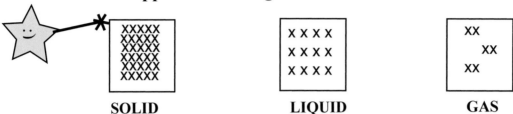

| SOLID | LIQUID | GAS |

Wish Star continued. "Matter in these three states is different from each other because of the closeness of their atoms. **Solids** hold their shape because their atoms are very close together. **Liquids** can be poured because there is more space between their atoms than solids, and they can flow over each other. The atoms of a **Gas** are so far apart that they can and do move in every direction."

"Guy, when I talked of **atoms** being close together in these examples, I was talking about **elements**. Elements are made of atoms. When you have matter that is a compound or a mixture, we call their smallest parts molecules. **If matter is not an element, we talk about them as space between their molecules.** There are some elements that are gases which like to travel in pairs like: Oxygen, Hydrogen, and Chlorine. We talk about these gases that have more than one atom traveling together as having space between their molecules."

Gas Molecules in Motion

Wish Star wanted to tell Guy some information about gases. Guy turned to listen. "Have you ever looked at a beam of sunlight and watched dust particles moving around?"

Guy said, "No, I haven't."

"There's a beam of sunlight over there between those two trees. Come follow me." Wish Star shook a dusty rag in the beam of sunlight and said, "Look Guy!"

Guy said, "I see the dust moving up, down, sideways, in a circle and every which way."

Wish Star said, "The dust particles are being hit by air molecules that are moving in every different direction. This is because the air molecules are so far apart that nothing prevents them from moving any which way. So the dust particles are showing us how molecules of air are moving—in every different direction."

"Amazing!" said Guy excitedly, "I didn't know that it was the air molecules hitting the dust particles that made them move so erratically. States of Matter explains so much."

The 4th State of Matter

"There is a **4th State of Matter. It's called Plasma**. Let's see what Comet Star's book has to say about it." One tap of Comet Star's book and out floated a whole paragraph of information in front of Guy.

Plasma

Plasma is found principally in the sun and other stars. Wish Star said, "Now you know why I just had to tell you about Plasma. I'm a star and when I'm out in space and being a real star. My usual state is plasma. Remember they turned off the real me so I could come to help you."

"Plasma is created because the temperature of the sun and other stars reaches thousands of degrees. When an atom gets heated that hot, the electrons that are around the nucleus fly off. This lets the nucleus float around free. When one nucleus bumps into another nucleus, they fuse together giving off the heat and light. This is where the sun and other stars get their heat and light. Earth needs this heat and light to support life. So this **4th State of Matter** is very important to the people here on earth."

The States of Matter Slide Back into Comet Star's Book

Wish Star said, "Now, you know about the States of Matter." So he tapped the book and away went the States of Matter. It's time to take a break Guy." Wish Star flew up to the top of the tree to rest. Guy went over to his blanket under a shade tree, and took a drink of cool water. As he drank it, he recognized that the water was a liquid in a solid paper cup, and the air around him was a gas. He was getting to know more and more about his world.

Guy's Atom Dream

He sat on his blanket resting against his soft backpack. It was nice and comfortable. He was sleepy. He thought about what he had learned concerning solids, liquids and gases. As he rested, he slipped into a dream. The whole world around him suddenly came to life displaying molecules. The tree pulled back his bark and showed Guy how close together the millions of molecules in his trunk really were. The tree called out to Guy, "Now you know why we are strong enough to stand so straight and tall. It's because our molecules are so close together. We are a solid." Looking over at the stream, Guy watched in amazement as the molecules of water called out to him, "Watch us tumble over each other as we flow by you. We're so glad to be liquid." A bird fluttered down before him, and said, "Watch the molecules of the air part as my wings flap, and I just glide right through the air because air's gas molecules are so far apart." After resting quite a while, Guy felt the fresh mountain air brush against his face, and he awoke from his dream.

Change of State

Wish Star flew down from the tree top and said, "Another contribution to the variety of matter in our world is this. Matter can change from one state to another. When it does, it still remains the same kind of matter it was before, but it looks different. Water is still water whether it is liquid water, the gas–water vapor, or the solid–ice. However, it looks entirely different in each state. Let's see what Comet Star has to say about **Change of State**." Wish Star tapped the book and out came a sentence.

Adding or Taking Away Heat Causes a Change of State.
Wish Star said, "Let's see what Comet Star's book tells us about Change of State." Wish Star tapped again and out came......

Adding Heat and Change of State
Comet Star's Book explains
Solids Changing to Liquids
MELTING

1. Frosty the Snowman
This question floated out of Comet Star's Book. **Do you remember what happened to Frosty the Snowman when the sun heated up the air around him?**

"Of course," said Guy, "He melted."

Comet Star' book explained, "The sun heated up the air around Frosty. He was made of snow, a solid. The sun's heat made the solid snow change into the liquid, water. So the sad story of Frosty melting away and becoming a puddle of water was an example of **Change of State**—A solid, the snow changing to a liquid, the puddle of water."

2. The Ice Cube
Guy said, "When an ice cube falls to the floor out of our ice dispenser, the heat of the room melts it, and you have a tiny puddle on the floor."

Wish Star said,"This is an ice cube changing to a puddle–a solid to a liquid."

3. The Ice Cream Cone
Guy said, "In the summer when I'm eating an ice cream cone, if I don't eat it fast it starts to melt. The solid ice cream turns to a liquid and flows down my arm. Then it gets me all messy. I guess that's a **Change of State** too. The solid ice cream becomes a liquid. I know its a liquid because liquids flow."

Liquids Changing to Gases
EVAPORATION
Out of Comet Star's book floated the words......
"If you add enough heat to a liquid, it can change into a gas.
Comet Star's book explains
Liquids Changing to a Gas

1. Water Boiling
Water boils on the stove. Water, a liquid, changes into water vapor, a gas. Water vapor is invisible. So what you see above the water is steam. This is the gas water vapor changing back into tiny drops of liquid water which are visible. Steam would not have happened if water, a liquid had not changed into water vapor, a gas.

2. Summer Puddles

After a summer rain storm there are puddles. The hot sun comes out, heats up the liquid puddles and they disappear. You don't actually see the liquid rain water become a gas because this happens so slowly. However, you can tell it happened. The puddle is gone, and you can feel that the water vapor is in the air. The air feels humid. The liquid puddle changed into the gas, water vapor, the humidity in the air. This is a liquid changing to a gas.

3. Bath Towel Drying

Guy had another example. "After my shower, I hang up my wet towel that dried me. The next day it is dry. and ready to use again. I guess the water in the towel got heated by the air in the room, and changed to water vapor. The water in my towel moved into the air in the house, and my towel got dry. That's a **Change of State** from a liquid to a gas. But I never thought about it. Oh, I'm learning so much!" said Guy.

Wish Star gave the book another tap. Out came the words

Taking Away Heat and Change of State

Comet Star's book explains

Liquids Changing to Solids

1. The Lake Freezes in Winter

When cold winter winds take away heat from the lake, the lake freezes over. This is a change of state from liquid water to a solid, ice.

2. Refrigerator Makes Ice Cubes

When the refrigerator takes heat out of the water in the ice cube trays, the liquid water changes to a solid—the ice cubes.

When we take away enough heat from any kind of liquid, it changes to a solid."

Comet Star's book explains

Gases Changing to Liquids

1. A Cold Drink in Summer

Have you ever observed an ice cold glass of soda on a hot summer day form drops of water on the outside of the glass? They say the glass is sweating. That's the water vapor in the air around the glass changing to a liquid on the outside of the glass. The cold drink took heat out of the air around the glass, and changed the water vapor, a gas, into a liquid—drops of water that run down the outside of your glass. This is change of state from a gas to a liquid.The water vapor—a gas in the air is cooled and turns into a liquid, drops of water on the glass.

2. Bathroom Mirror Fogs Up

Guy said, "When I take a very hot shower, the bathroom mirror fogs up. The shower puts water vapor into the air. When it hits the cold mirror the gas, water vapor, changes to a liquid, tiny droplets of water, on the mirror. A gas changes into a liquid."

Wish Star's Quiz for Guy

1. "In the winter a car's windshield freezes over. Your father turns on the car's heater and the frost on the windshield melts. What kind of a **Change of State** is this?" _____ changes to a _____ .

Guy explained, "The frost is a **solid.** Heat changed it into a **liquid** drops of water. Answer: <u>solid</u> changes to a <u>liquid."</u>

2. Wish Star offered this story. "It's a hot humid summer day. A lady with eye glasses is in a store with freezing cold air conditioning. She walks out of the store into the hot humid outdoors with her very cold eye glasses, her glasses fog over. The change of state was:_____ changes to a _____ .

Guy said, "On a humid day, there is the gas, water vapor in the air. The lady's glasses are very cold, the water vapor in the air is cooled and changes into a liquid and fogs up the glasses.   Answer: <u>gas</u> changes to a <u>liquid</u>."

With a tap of his magic wand, Wish Star sent all this good information about **Change of State** back into Comet Star's book. Then he turned to Guy and said, "I guess you can see how Change of State is related to your original question about the great variety in matter. Change of State is just another way of increasing the variety of things in our world. Ice, water, and steam look entirely different yet they are all water, made of the same two elements, Hydrogen and Oxygen. "

"There are a thousand more examples of how the basic 92 elements can change and make the matter in our world look so different. Most of these changes are caused by elements combing and forming compounds. New matter is created that is entirely different from the elements that combine. Chemistry will explain so much of the world around you. You will love how much you are going to learn."

Forms of Matter

Professor Terry returned from her meeting. Wish Star flew down from the treetop where he had decided to rest a while. He shared with Professor Terry what Guy had learned while she was away.

Professor Terry said, "Before the end of the day, you are going to learn about the *Forms of Matter.* Wish Star, we need your magic wand." One tap and floating out of the book came the words………………....

Pure Substances: Elements and Compounds

These words swirled around in front of Guy, and he loved it. Then another word floated out of Comet Stars magical book.

Mixtures

"Yes Guy, before you go back to your cabin today, you are going to know what makes matter an element, a compound or a mixture. Comet Star's book will tell us what they are. Then we are going on a scavenger hunt to find samples of them here on the mountain. Come on Wish Star tap your magic wand on the book. We need to know what these kinds of chemicals are."

 **Tap 1** Out came the definition of a **Pure Substance**..........

A *Pure Substance* is matter that has a definite composition.

There are two kinds of *pure substances:* elements and compounds. If a sample contains only one kind of atom, the pure substance is an element. If the sample contains two or more kinds of *atoms chemically combined* then the sample would be *a compound. Elements and compounds are pure substances.* If a sample is not the same throughout, it would not be a pure substance. It would be a *mixture.*

 **Tap 2** Wish Star tapped with his magic wand and out floated the reason **Why Elements are Pure Substances:**

***Elements* are pure substances because they are made up of only one kind of atom.**

Hydrogen gas is a pure substance when the sample is made up of only one kind of atom, Hydrogen atoms. Lead is a pure substance if the sample is made up of only one kind of atom, Lead atoms. Copper would be a pure substance if it contained only atoms of copper.

Professor Terry turned to Guy and said, "People send me requests often to find out if what they are buying is really pure gold or pure silver. I test them and find out if I can only detect atoms of gold in the gold sample or atoms of silver in the silver sample. If I find that there are other elements or compounds mixed in then it would not be a sample of a pure substance

 **Tap 3** Wish Star was having fun tapping the book now. He started to do it with a fanfare twirling around and being silly. Out of the book floated the reason compounds are pure substances:

***Compounds* are pure substances because they are made up of a definite number of different atoms chemically combined.**

The compound, water is a pure substance because it is made up of exactly two atoms of Hydrogen and one atom of Oxygen chemically combined, H_2O. The compound, carbon dioxide is a pure substance because it's always made up of one atom of Carbon and 2 atoms of Oxygen chemically combined, CO_2.

 **Tap 4** Wish Star was really enjoying using his magic wand. This time he twirled around and bent at the waist before he did Tap 4. Out floated the word mixtures and its definition:

A *Mixture* is a sample of matter that has two or more substances mingled together, <u>not</u> chemically combined, and easily separated.

Professor Terry said, "As an example, we can make a mixture by stirring the compound sugar into the compound water. The sugar mingles with the water, and you can't see it any more. Here two compounds mingled together. It is not just one compound, so it is not a pure substance. It is a mixture.The two do not chemically combine and do not form anything different. If we wish, we can separate the water from the sugar by simply evaporating the water. When the water is gone, the sugar that had been mixed with the water is left. If the sugar and water had chemically combined, you would not be able to get the sugar back by such a simple means."

Wish Star's Work is Done
Professor Terry Promises to Teach Guy

Professor Terry said, "Wish Star, what would we have done without you! Now Guy knows all the basic vocabulary that he needs to continue learning chemistry."

"Guy, I guess my work is done," Wish Star said sadly as he had grown fond of the young boy. "Comet Star sent me here to assure that you would get your wish to learn chemistry. Professor Terry will take over from here, and make sure you learn all that basic chemistry can teach you about your magnificent world. You got your wish. I must go now, but I'll be watching my little friend from on high. I won't forget you."

Guy said, "Thanks, Wish Star, it was fun getting to know you. You made my wish come true. I'll never forget you."

Then Wish Star turned and zoomed in an arc over the trees. He paused and winked at Guy before he disappeared behind the clouds into the sky above.

Forms of Matter on the Mountain

Professor Terry said, "We're going to really miss Wish Star. Now it's my job to help you understand your world. Let's see what chemistry we can find right here in this beautiful tree lined grassy field. How about looking for elements, compounds and mixtures right here?"

Guy said, "What a fun afternoon—chemistry right here on my mountain!"

Are Guy's Coins Pure Substances?

Guy turned to Professor Terry and said, "Here are some coins I found by the stream. Wish Star's book told us about elements. I think my coins are samples of elements, which are pure substances. However I'm not sure the coins are *only* made up of the atoms of these elements. They may be mixtures. One coin looks like my family's copper kettle. Copper is an element. So maybe it's a pure substance with nothing else mixed in. If it is, in every part of the coin there will only be atoms of Copper. The next coin is Silver. If all the parts of this coin have only Silver atoms, then it is an example of a pure substance, too. The last coin looks like Gold. If all parts of it are atoms of Gold, then it would be an example of a pure substance, an element."

Professor Terry said, "I really doubt that this coin is pure Gold. A coin that big, if it was pure Gold, would be worth a great deal of money. It would not have been left carelessly by the stream. Anyway, I will test all your coins in the lab. Then we will know if the coins are pure substances. If they are mixtures, I will find other matter mixed in with the elements that these coins appear to be."

Looking at the coins, Guy wondered if they were the pure substances, elements. Or were they mixtures of elements and compounds.

A Search for a Compound

Professor Terry continued, "Compounds are pure substances. Guy, let's look around and see if we can find any examples of compounds. Remember, what Wish Star's book said about compounds: **Compounds are pure substances made up of two or more elements chemically combined.**"

"Let me give you an example. Two gases, Hydrogen and Oxygen chemically combine and form something new and different – the compound water. Water, a liquid, is very different from the gas elements from which it was made. Water is a compound, a pure substance. This stream is made up of the compound water but the stream is not a pure substance You can see other things mixed in the water. It's a mixture of the compound water, a pure substance, and mud and leaves."

Guy said, "I see the mud dissolved in it. So I understand why the stream is not a sample of a pure substance. It's a mixture." Guy had a puzzled look and said, "I don't know what chemically combined means."

"That means they are joined together in a strong way and not easy to take apart," Professor Terry responded. "Water is not broken down easily into the two gases that made it. There's a lot more to learn about compounds, but for now that's all I'll tell you. Learning what chemically combined means is something I've planned to teach you in a special way, in a special place."

Professor Terry's response had a hint of mystery about it. Guy wondered what she meant, as he eagerly followed her toward the stream. Professor Terry found a rusty nail beside one of the rocks. "Guy look at this rusty nail I found. A nail is made of the element Iron. When it rusts it combines chemically with Oxygen in the air to form a compound, Iron oxide that is rust. Rust is a compound made up of the elements Iron and Oxygen. Two atoms of Iron chemically combine with three atoms of Oxygen. They are no longer the elements Iron and Oxygen. They are now something new, a compound with a definite composition, Iron Oxide. The compound looks very different from the elements that made it. Iron a hard silver colored metal and oxygen a colorless gas. Rust's color is now red and it's not hard any more. It is something new. Rust is a pure substance called a compound made up or 2 elements chemically combined."

"We have found what could be a pure substance–elements in your coins. We have found a compound, the rusty nail. It was a sample of the pure substance, a compound. We observed that the water in the stream was an example of a mixture. Now, let's find other examples of mixtures. Let's start looking."

Guy Searches for Mixtures with Professor Terry

Professor Terry helped Guy recall what a mixture was like. **"A mixture is what you get when you mingle two or more substances together, but they do not combine chemically.** The substances mix together but don't bond. That makes them a mixture."

Guy asked, "How you can tell if substances bond or if they don't?"

Professor Terry answered, "They are not bonded if you can separate them easily using simple physical means: like using a magnet, pouring them through a filter or a sieve. Chemists use evaporation, distillation and many other methods to physically separate parts of a mixture and get back the original matter that was mixed together. Now I'm getting too far ahead. You'll learn about these methods scientists use later. We can find many examples of mixtures around here if we look. Over here, Guy! Let's scoop up some of this soil and look at it."

Guy observed, "There's some brown sandy soil with some crushed leaves and a few pebbles mingled in it. This is a mixture."

Professor Terry affirmed Guy's assessment that this soil sample was a mixture. "You were right when you called it a mixture because it can be easily separated. We could put this soil sample through a sieve. The sandy soil will go through easily, and the crushed leaves and pebbles will remain in the sieve. Then we can pick out the few pebbles that are in with the leaves. The parts were obviously not combined and were easily separated by the physical means of sifting and sorting. This is obviously a mixture. You can see three types of matter mingled together and they never combined."

Professor Terry continued adding more information about the soil mixture. "It's called a **heterogeneous mixture** because we can see its different parts. The example I gave you of sugar mixed with water is a **homogeneous mixture** because you can't see the sugar once it dissolves in the water."

Guy said, "I sure have learned a lot about mixtures."

Guy Performs a Mixture Experiment

They got up and walked back to Guy's blanket. Spying Guy's lunch bucket, Professor Terry said, "What a good mother you have sending a lunch with you! Let's see what you have for lunch. Looks like she made a great lunch for you. In addition to the good food, she provided you with several things we can use to create a mixture. I see a bottle of water, a packet of salt, and a small paper cup."

She said, "Guy, I'm going to let you do an experiment. Mix some salt with a small amount of water in this little cup. Swish it around to make the water and salt mingle well. Salt and water are compounds. So, we are trying to make a mixture of two compounds. The salt will mix so evenly throughout the water that you will not see it anymore. That will make it a homogeneous mixture. The soil, leaves and pebble mixture we found before was called a heterogeneous mixture because in that mixture you could see all its separate parts."

Guy was happy to be experimenting like a scientist. After he mixed the salt and water, he gave Professor Terry the paper cup with the salt and water mixture in it.

Professor Terry took it and suggested, "Let's leave it in the hot sun, and the water will evaporate. If this is a mixture, the salt will not chemically combine with water. When the water evaporates, the same salt you mixed into the water will be left in the cup. That will be proof you made a mixture. Let's see if you will get the salt back tomorrow."

"Guy, let's move the cup into a sunnier place where it will not get knocked over. Tomorrow you can check if the salt reappears in the cup after the water has evaporated. If it does, you will know you created a mixture."

"Now, I'll tell you about a few other examples of mixtures. Take a deep breath, Guy. The air you just breathed in is a mixture of several gases: Oxygen, Nitrogen, and Carbon Dioxide as well as small amounts of other gases. The ocean is a mixture. It contains the compound water mixed with different kinds of salt compounds. When you study chemistry in school, there will be a whole section dedicated to mixtures. You will learn about solutions—unsaturated, saturated, and supersaturated. You will learn about the parts of a solution: the solute and the solvent. You will love learning about them. Chemistry has so much to teach you."

"Wow! Professor," Guy said, "You've really helped me expand my knowledge of my world."

"Chemistry has a vocabulary of its own, Guy. So it's important to learn this vocabulary before we go on. Let's see how much of this scientific vocabulary you remember. I have a quiz for you to see how much you have learned."

"When you finish the quiz, you can check your answers. If you get any wrong, go back and take the quiz over. Repeat taking the quiz, as many times as necessary, until you know what each scientific word means without looking at the answer sheet."

Professor Terry slipped this answer sheet into Guy's folder saying, "Don't look at the answers until you finish the quiz. Then check to see which ones you need to review.

Guy's Quiz on The Basic Vocabulary of Chemistry

1. Anything that occupies space and has mass is _____
2. The amount of matter in a body. _____
3. The pull of gravity on the mass of a body, _____
4. The basic unit of matter in the universe_____
5. The parts of the atom are the _____ and the _____
6. The particles in the energy levels of the atom are _____
7. The particles in the nucleus of the atom are the _____ and the_____
8. Na, Cl, K, C, Mg and Al are _____ for the elements.
9. The 118 elements are listed on The _____ _____
10.These are pure substances._____ and _____
11._____ is the fourth state of matter.
12. The study of matter is called _____
13. The 3 states of matter are_____
14. Two or more substances mingled together without bonding is called a _____
15. A substance made up of two or more elements chemically combined is a_____
16. A pure substance with only one kind of atom in it is called an _____
17. When you add enough heat, a solid becomes a _____
18. If you add enough heat, a liquid becomes a _____
19. If you take away enough heat, a liquid becomes a _____
20. If you take away enough heat a gas becomes a _____

Professor Terry suggested "Take out the answer sheet and check your answers. It will help you to learn the questions you got wrong. Then study that vocabulary more.

Guy's Folder
Answers to the Quiz on The Basic Vocabulary of Chemistry
1. matter 2. mass 3. weight 4. atom 5. nucleus and energy levels 6. electrons
7. neutrons and protons 8. symbols 9. The Periodic Table 10. elements and compounds
11.plasma 12. Chemistry 13. solids, liquids, gases 14. mixture 15. compound
16. element 17. liquid 18. gas 19. solid 20. liquid

"You need to remember this important vocabulary. A good way to do this is is to turn these questions into a word game. Create your own game like one of those TV game shows. Playing games is a good way to make sure you remember this vocabulary"

Guy liked this idea. He could get his cousins to play with him when they visit.

"Next, I'll teach you what the Periodic Table has to do with the Atom. My Periodic Table is unique. You will find it *most interesting*." She left saying, "See you Monday!"

As Guy made his way up the mountain to his parents' cabin, he began to wondered what could be *most interesting* about her Periodic Table. It was not so much what she said. It was her mysterious tone of voice that intrigued him. He pondered the question, "What could be so different about her Periodic Table *?"*

Guy will find out soon enough. What a great summer he has ahead of him!

BOOK 1
PART 2
Guy in Periodic Table Land

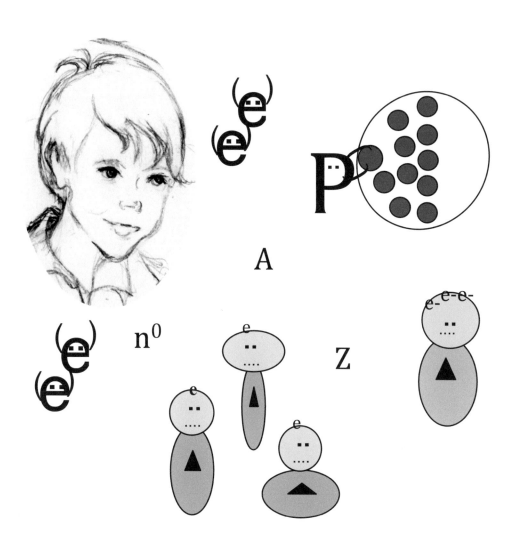

Irene P. Reisinger

BOOK 1
PART 2
Table of Contents

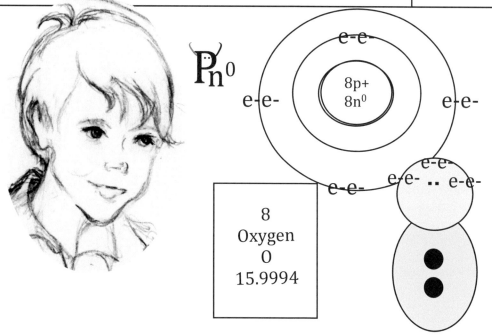

P_n^0

8p+
8n0

8
Oxygen
O
15.9994

Irene P. Reisinger

PART 2
Chapter 1
Guy's Adventure in Periodic Table Land Begins

On Monday as Guy headed to the university lab, the sun was shining just above the mountain tops. There was a sweet smell of honeysuckle in the air, and Guy bent over to admire the blue cornflowers blooming beside the road. He loved his mountain and all its beautiful wildflowers. He also loved what he was learning about his world. He eagerly continued down the mountain thinking about what could make Professor Terry's Periodic Table so interesting?

When Guy entered the chemistry lab on the university campus, a large chart on the back wall caught his attention. The title of the chart **The Periodic Table of the Elements** seemed to jump out at him. This has to be the chart that Professor Terry was talking about, thought Guy, as he walked in between and around the slate topped lab tables to get a closer look. I see the symbols that I memorized for elements, Atomic Numbers 1 to 20. She told me I'd find them on this chart. I wonder what other information this chart holds for me to learn.

Professor Terry came out of the storage room with a tray full of glassware and spotted Guy looking at the chart. She placed the beakers and flasks down on the lab table

and greeted him. "It's good to see you, Guy. I'm glad you're here. I've got so much to teach you today."

Guy made his way around the lab tables to get closer to her and said, "Hi Professor Terry, I saw that chart on the back wall and felt it had to be the one you were telling me about. I never expected it to be so big."

"That's the chart all right, and I made a smaller version of it for you to put in your folder with the Element/Symbol paper. Soon you will be learning all about the Periodic Table," said Professor Terry.

"I memorized the first twenty elements with their symbols over the weekend, and I found them on the chart. You can be proud of me. I worked hard to learn all those symbols for the elements."

Guy's Goal –Understanding the Periodic Table

"I am proud of you," said Professor Terry. "It makes me so happy when I find young children like you excited to learn about our world. I already told you that everything in our entire universe is made up of atoms of the elements. Just think, they are all listed on this Periodic Table. So learning the symbols for the first twenty of those elements was the perfect place to start. Now let's begin discovering what else the Periodic Table has to do with the elements whose symbols you memorized. You will find all the elements ever discovered on this chart with their symbols and atomic numbers."

"Each of the 118 elements you see on the Periodic Table has an atom that is distinctly different from all the others. The principal difference between elements is that the atom of each different element has a different number of protons in its nucleus. It's the number of these protons in the element's atom that determines its Atomic Number. That's what makes that element different from all others. Every atom that has the same number of protons in its nucleus is that exact same element. Sodium has 11 protons in its nucleus. Gold has 79 protons in its nucleus. No other element has those exact number of protons. If an atom has 11 protons it is Sodium. If an atom has 79 protons in its nucleus it is Gold. You can tell how many protons the different elements have by looking at the Periodic Table. The number of protons is equal to the element's Atomic Number. You find that number at the top of each element box."

"You can also use the Periodic Table to figure out the number of electrons and neutrons that each of the elements should have in its atom. The element box on the Periodic Table is the key to figuring out *how many* of these particles are in each atom. Being able to use the Periodic Table to **figure out the number of particles in each atom is your first major goal.** When you understand the Periodic Table well enough, you will be able to draw diagrams of these atoms. These diagrams are called Bohr models. Drawing Bohr models is your ultimate goal. Guy, a goal is something you reach by taking little steps one after another until you know enough to reach it. Your goal, drawing Bohr models, will be reached when you have learned to use the Periodic Table. That's going to take many small steps to achieve, but in the end you will know how to draw Bohr models. The Periodic Table is the key to determining the number of electrons, protons, and neutrons in the atom. This is what we will be working on to make your goal possible and we will start today. Let me tell you a little bit of information about Bohr models."

"Bohr models are named after the scientist who developed them, Niels Bohr. There are many different models of the atom, but the Bohr model is the most useful in learning how these elements form compounds and engage in chemical reactions. Before the summer is over you will be learning about these chemical principles. Now you are about to begin your study of the Periodic Table. Let's get started, Guy."

Guy Learns About Professor Terry's Magical Periodic Table

Professor Terry began, " I have a secret to share with you. My Periodic Table is special. It is different from any other Periodic Table. Well, I have to tell you, it's more than different—it's magical! It is going to play an important role in helping you to learn chemistry. I can't wait to share with you how exciting it will be to learn chemistry this way. I should say *in this place*. But before I introduce you to the magical side of my table, I'd like you to learn several things about my Periodic Table while it is still in its normal state."

Guy thought, that he was right yesterday when he believed Professor Terry was hinting that her Periodic Table was mysterious. Now she confirmed his suspicion.

"Let's move to the back of the lab near the Periodic Table you spotted when you first arrived. I will begin by having you pay special attention to two items that are on every Periodic Table—**the Groups and the Periods**. They are found on the top left corner of the Periodic Table.

Guy Learns About the Groups on the Periodic Table

Professor Terry first pointed to the top left corner of the Periodic Table where it had arrows pointing to where you could find the Groups and the Periods. She said, "We will begin with the Groups first." She took her laser pointer and the little red dot soon skipped over the tops of each column on the Periodic Table where the Group arrow was pointing.

PERIODS
GROUPS→ **The Groups: 1A, 2A B Groups 3A, 4A, 5A, 6A, 7A and 8A.**

"These are the Groups," she announced. "Notice, there are A and B Groups. I will be explaining the rules that apply to the elements in the A Groups. Since the B Group elements do not follow these rules, you'll learn about B Groups later."

Then Professor Terry ran her pointer across the top of the Periodic Table again. This time she drew attention only to the A Groups. She skipped over the B Groups in the middle of the chart.

"The A Groups tell how many electrons are in the Outside Energy Level of every element in its column. All the elements in Group 1A have one electron in the Outside Energy Level. All the elements in Group 2A have 2 electrons, 3A have 3; 4A, 4. If you know the Group number, you know the number of electrons in the element's Outside Energy Level."

Professor Terry went over to the blackboard and drew two atoms. Guy watched and wondered what she was doing. Then Professor Terry explained . "I've drawn 2 atoms

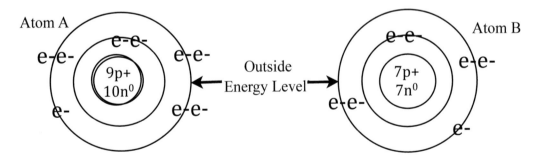

so you can see what I mean about the Group Numbers, and the Outside Energy Levels.." Guy said, "I guess what you want me to do is to count the number of electrons in the Outside Energy Levels of these 2 atoms." Guy went up to the board and did just that. **Atom A had 7 electrons in its Outside Energy Level and so it was in Group 7A. Atom B had 5 electrons in it's Outside Energy Level. So it was in Group 5A.**

Professor Terry said, "That's right. You counted the number of electrons in the energy level furthest away from the nucleus. Then you put an A after that number, and you had the Group Number for that atom. Good job, Guy!"

"You now know what the Group Number means. If you count the electrons in the Outside Energy Level of an atom you can tell which Group the element is in. Then you can go to the Periodic Table and find that Group Number. The element will be in that column."

Guy Learns About the Periods

PERIODS

1.Again, Professor Terry pointed to the top left hand corner of the Periodic Table to the
2. arrows which indicated where you would find the Groups and the Periods.
3.Then she slid the red dot of her laser pointer down the left side of the chart over the
4.numbers 1 to 7. She told Guy, "Take a look at the Periodic Table in the back of my lab..
5.Notice that these numbers are in front of each row of elements. **All the**
6.**elements in that row have the same number of energy levels indicated by the**
7.**Period number."** She then told Guy, "Look at the Periodic Table and find Period 2 down the left side.

Guy found Period 2 and here are the symbols for the elements in that Period.

Period 2 Li, Be……………………………………………….. B, C, N, O, F Ne

"These are symbols for some of the elements that I memorized." Guy was excited to recognize something familiar. I remember most of them. However I get the symbols Be, Beryllium and B, Boron mixed up. I have the same trouble with N, Nitrogen and Ne, Neon. At least I find it easy to remember the rest of the elements in Period 2: Lithium, Carbon, Oxygen and Fluorine."

Professor Terry said, "Hint, Guy. The symbol Be couldn't be Boron. and Ne could not be Nitrogen. Why?"

Guy remembered, "Because the second letter in the symbol is the second letter in the element's name (except for Chlorine and Magnesium). I guess from now on I'll remember Be is **Be**ryllium, and Ne has to be **Ne**on."

Professor Terry directed Guy to look at Period 3 on the Periodic Table. Then she asked, "What are the first two elements in Period 3?"

Guy looked at the top left corner of the Periodic, and found the word Period next to the word Group. Then he followed the arrow from the word Period down the left side of the Periodic Table. There was Period 3 right after Period 2. He then looked at the first two elements and found that they were Sodium, Na and Magnesium, Mg.

Period 3:**Na, Mg**……..………………………………..Al, Si, P, S, Cl, Ar

Professor Terry reviewed the meaning of the Period numbers one more time. "**The Periods tell us how many energy levels are in an atom.** All the elements in the Period 3 have 3 energy levels. All the elements in Period 5 have 5 energy levels."

"Guy, look at the atoms A and B that I drew to teach you the meaning of the Group Numbers. Tell me what Period they are in. One word of caution. **Don't count the**

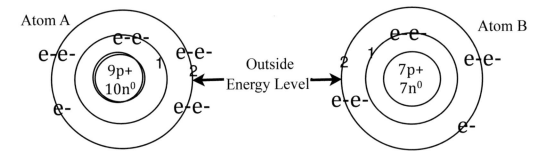

circle for the nucleus. Notice the energy levels all have electrons in them."

Guy counted the energy levels and said, "Both Atom A and Atom B have 2 energy levels. So they are both in Period 2 on the Periodic Table."

"Now you know the meaning of Periods on the Periodic Table. Both the Groups and the Periods will be significant in helping you learn the number of particles in the atom of an element. You will need to know this to finally be able to draw Bohr models. This is what we will be working on now. This is what you will learn as you study the Periodic Table."

"One last question Guy. If an atom has 2 electrons in it's Outside Energy Level and 4 energy levels, tell me its Group Number and what Period it is in."

Guy responded, "The electrons in the Outside Energy Level tell us the Group Number. So it's in Group 2A. The number of energy levels in the atom tells us the Period Number. So it's Period 4. My answer is ……..Group 2A, Period 4."

Professor Terry was happy that Guy understood about the Groups and the Periods."

Special Features of Professor Terry's Periodic Table

Professor Terry said, "Now, you're ready to look at my Periodic Table. Besides being magical, my Periodic Table performs better than ordinary ones in its normal state. It lights up when I push these buttons." She pushed one button and the whole table lit up. Guy saw lots of red boxes, and yellow boxes, and a few orange ones "First, I will show you which of these elements are **metals** and which are **non-metals."** More than half the element boxes were red, indicating they were metals.

"I needed you to see this. Isn't it amazing how many elements are metals?" Professor Terry pointed out.

That's exactly what Guy was thinking. He was really surprised that more than half the elements were **metals**. As he looked around the lab, few things were metal."

Professor Terry explained, "In chemistry the meaning of the word **metal** goes far beyond what most people think of as a **metal**. People think of metals as hard and sometimes shiny. I will explain the chemical meaning of the word **metal** when you are ready to use this knowledge. You will learn that it is more about how the metal behaves

THE PERIODIC TABLE OF THE ELEMENTS

chemically than what it looks like. It's meaning is a fascinating part of how these elements listed on the Periodic Table become something very different called compounds."

"The elements that turned yellow are **non-metals**. They are mostly in Groups 5A, 6A, 7A and 8A. Yet Hydrogen is in Group 1A on the other side of the chart, far from all the other non-metals. Besides being far away from the other non-metals, it's located in a Group made up of all metals. You will learn why later. There's so much you have to learn to understand these facts."

Suddenly the orange elements started to blink. "Scientists call these elements **metalloids. Metalloids sometimes behave like metals and at other times they behave like non-metals.**"

As she finished explaining about metals, non-metals and metalloids, Professor Terry shut off these lights and said, "Guy, I have another button here which I'll let you push this time. It will bring up another Periodic Table. It will show the **States of Matter** of the elements." Guy pushed that button and the whole chart lit up again. Most of the

THE PERIODIC TABLE OF THE ELEMENTS

PERIODS ↓ GROUPS →

STATES OF MATTER
Solids- red
Liquids - Orange
Gases- Yellow

	1A	2A	3B	4B	5B	6B	7B		8B		1B	2B	3A	4A	5A	6A	7A	8A
1	1 H 1.00																	2 He 5..50
2	3 Li 6.94	4 Be 4.01											5 B 10.8	6 C 12.01	7 N 14.01	8 O 16.99	9 F 18.99	10 Ne 20.18
3	11 Na 22.9	12 Mg 24.3											13 Al 26.96	14 Si 28.09	15 P 30.97	16 S 32.07	17 Cl 35.45	18 Ar 39.95
4	19 K 39.1	20 Ca 40.08	21 Sc 44.95	22 Ti 47.86	23 V 50.94	24 Cr 51.99	25 Mn 54.93	26 Fe 56.84	27 Co 58.93	28 Ni 58.69	29 Cu 63.54	30 Zn 65.38	31 Ga 69.22	32 Ge 72.64	33 As 74.92	34 Se 78.96	35 Br 79.90	36 Kr 83.79
5	37 Rb 85.46	38 Sr 87.62	39 Y 88.90	40 Zr 91.22	41 Nb 92.90	42 Mo 95.96	43 Tc {98}	44 Ru 101.07	45 Rh 102.90	46 Pd 106.42	47 Ag 107.86	48 Cd 112.41	49 In 114.81	50 Sn 118.71	51 Sb 121.76	52 Te 126.90	53 I 126.90	54 Xe 131.29
6	55 Cs 132.91	56 Ba 137.32	57-71	72 Hf 178.40	73 Ta 180.94	74 W 183.84	75 Re 186.20	76 Os 190.23	77 Ir 192.21	78 Pt 195.08	79 Au 196.96	80 Hg 200.59	81 Tl 204.38	82 Pb 207.4	83 Bi 208.98	84 Po {209}	85 At {210}	86 Rn {222}
7	87 Fr {223}	88 Ra {226}	89-103	104 Rf {267}	105 Db {268}	106 Sg {271}	107 Bh {272}	108 Hs {270}	109 Mt {278}	110 Gs {281}	111 Rg {280}	112 Cn {285}	113 Uut {284}	114 Fl {289}	115 Uup {288}	116 Lv {203}	117 Uus {294}	118 Uuo {294}

57 La 138.90	58 Ce 140.11	59 Pr 140.90	60 Nd 144.24	61 Pm {145}	62 Sm 150.36	63 Eu 151.96	64 Gd 157.35	65 Tb 158.92	66 Dy 162.50	67 Ho 164.95	68 Er 167.25	69 Tm 168.93	70 Yb 173.06	71 Lu 174.96
89 Ac {227}	90 Th 232.03	91 Pa 231.03	92 U 238.02	93 Np {237}	94 Pu {234}	95 Am {243}	96 Cm {247}	97 Bk {247}	98 Cf {251}	99 Es {252}	100 Fm {257}	101 Md {258}	102 No {259}	103 Lr {262}

elements were red. Guy was not surprised that almost all the metals proved to be solids. Guy had no opinion about metalloids but they turned red too.

Professor Terry added, "Even some non-metals are solids. What is surprising is that there are only two elements that are liquids. Notice, which two elements turned orange."

"The world is so full of liquids," said Guy. "How can this be?"

Professor Terry explained, "Remember, this chart only lists elements. Obviously, most liquids we know must be compounds or mixtures."

Guy said, "I'm glad you explained that. Now, I can understand how it is possible, that only two elements are liquids."

"What is really surprising," continued Professor Terry, "is that one of the two elements that are liquids is a metal. We always think of metals as solids. However, Mercury is a liquid, but like so many metals, it is shiny and silvery. Perhaps you have seen Mercury in thermometers. It's that silvery metal that expands when heated and shows whether or not you have a fever. It's not used much any more because Mercury is harmful to your health if the thermometer breaks. The other liquid element is Bromine. It's not well known, so there is no reason to be surprised about it being a liquid."

"Notice the boxes that lighted up yellow," said Professor Terry. "They are the elements that are gases: Hydrogen, Nitrogen, Oxygen, Fluorine, Chlorine, Helium, Neon, Argon, Krypton, Xenon and Radon. If you remember, all of these elements were non-metals. Well, these are the most important facts I wanted to teach you about the Periodic Table while it still looks like a normal table."

Guy's curiosity grew when she uttered the words, *while it looks like a normal table.* He wondered, "Is the Periodic Table going to change in some mysterious way?"

"You now know where to look for the Groups and Periods. I'm giving you copies of these two Periodic Tables. One shows the elements that are solids, liquids and gases. The other shows the elements that are metals, non-metals and metalloids. This is for you to keep with your important papers, and we can refer to them later. What I have shown you just now will become more understandable as you get deeper into the study of chemistry."

"Before you leave, Guy, I want you to see the magical side of my Periodic Table that will be a big part of your learning experience this summer."

Guy Goes to Periodic Table Land

As they moved closer to the Periodic Table, Professor Terry assured Guy, "What we are about to experience is entirely safe and it will be fantastically wonderful. Come now Guy lets lean against the Periodic Table and hold my hand." They both leaned on it and the whole section of the wall pushed in and flipped around. Bells, whistles and sirens screamed—adventure ahead. Professor Terry said, "Here we go Guy!" She held Guy's hand tighter and they found themselves sliding down a tunnel of mirrors reflecting a thousand sparkling lights. Magical sprinkle dust enveloped them like a silky blanket. They felt comforted by soft mysterious music. Then, in an instant, they were in a strange new world. Guy couldn't believe his eyes. He was greeted by colorful blinking lights surrounding all the street signs. All around were adorable houses of an amazing variety of shapes and sizes. Guy soon realized he was in another world, definitely a fantasy world.

Professor Terry said, "This is Periodic Table Land."

Guy pointed to the street signs surrounded by colorful, blinking lights and questioned, "Weren't those street signs the Group numbers on the Periodic Table?"

Professor Terry affirmed his observation. "Yes, Guy. What was 1A is now First Street, 2A, Second Street. All the B Groups have become B Avenue. After that you'll seeThird, Fourth, Fifth, Sixth, Seventh and Eighth Streets which used to be Groups 3A, 4A, 5A, 6A, 7A, and 8A." Professor Terry added, "The houses are what were the element boxes on the Periodic Table. Each element has his own house."

Then Guy glanced down the left side of the town where there were many different colored lights on short poles with numbers on them. Professor Terry, noticing where Guy had directed his eyes, said, "Yes, they are what had been the Period numbers on my Periodic Table."

Professor Terry stood there as Guy gazed in amazement at the element boxes now turned into the cutest, most interesting and unique houses anyone has ever seen. Some of the elements came out of their houses and were walking here, there and everywhere. Guy was spellbound. He did not know what to say. He was amazed as he looked all around at

Irene P. Reisinger

the elements' houses. The colorful blinking lights lent an aura of festivity to this fantastic world.

Professor Terry said, "In the next couple of days I will arrange for you to tour Periodic Table Land. You will meet a few of the elements. This summer you will spend much time here learning everything you need to know about the Periodic Table."

They stayed for a short time and then slipped back into the lab and the real world.

Back in the Lab

When back in the lab, Professor Terry said she had a few reports to write up. She suggested Guy take out his Element/Symbol chart and review the symbols. Guy searched

Professor Terry's Element/Symbol Chart
Atomic Number/ Name of Element/ Symbol

1	Hydrogen	H	8	Oxygen	O	15	Phosphorus	P
2	Helium	He	9	Fluorine	F	16	Sulfur	S
3	Lithium	Li	10	Neon	Ne	17	Chlorine	Cl
4	Beryllium	Be	11	Sodium	Na	18	Argon	Ar
5	Boron	B	12	Magnesium	Mg	19	Potassium	K
6	Carbon	C	13	Aluminum	Al	20	Calcium	Ca
7	Nitrogen	N	14	Silicon	Si			

in his back-pack and found it. He didn't have a chance to look at it before Professor Terry returned with another chart in her hand.

She said,"I have a new chart for you. You probably noticed the boxes on the Periodic Tables I gave you were too small to write in the names of the elements. All you have now in the element boxes on your Periodic Table are the atomic number, the symbol and the atomic mass unit, the (amu). Soon you are going to need to search the Periodic Table given only the name of the element. So you are going to need a chart showing the atomic numbers of the elements to be able to find them on your Periodic Table."

"You have the names of elements atomic numbers 1 to 20 on your Element/ Symbol chart. This new chart gives you the atomic numbers and the names of elements 21 to 118 Keep both charts together with your Periodic Table. These charts will be used often. So keep them handy. Knowing the atomic number of an element, you will be able to locate any element you are asked to find on your Periodic Table."

Here's the chart. Look it over and find elements that are familiar to you. Pick out

ELEMENTS ATOMIC NUMBERS 21 TO 118

21	Scandium	47	Silver	73	Tantalum	99	Einsteinium
22	Titanium	48	Cadmium	74	Tungsten	100	Fermium
23	Vanadium	49	Indium	75	Rhenium	101	Mendelevium
24	Chromium	50	Tin	76	Osmium	102	Nobelium
25	Manganese	51	Antimony	77	Iridium	103	Lawrencium
26	Iron	52	Tellurium	78	Platinum	104	Rutherfordium
27	Cobalt	53	Iodine	79	Gold	105	Dubnium
28	Nickel	54	Xenon	80	Mercury	106	Seaborgium
29	Copper	55	Cesium	81	Thallium	107	Bohrium
30	Zinc	56	Radium	82	Lead	108	Hassium
31	Gallium	57	Lanthanum	83	Bismuth	109	Meitnerium
32	Germanium	58	Cerium	84	Polonium	110	Darmstadtium
33	Arsenic	59	Praseodymium	85	Astatine	111	Roentgenium
34	Selenium	60	Neodymium	86	Radon	112	Copernicium
35	Chlorine	61	Promethium	87	Francium	113	Nihonium
36	Argon	62	Samarium	88	Radium	114	Flerovium
37	Rubidium	63	Europium	89	Actinium	115	Moscovium
38	Strontium	64	Gadolinium	90	Thorium	116	Livermorium
39	Yttrium	65	Terbium	91	Protactinium	117	Tennessine
40	Zirconium	66	Dysprosium	92	Uranium	118	Oganesson
41	Niobium	67	Holmium	93	Neptunium		
42	Molybdenum	68	Erbium	94	Plutonium		
43	Technetium	69	Thulium	95	Americium		
44	Ruthenium	70	Ytterbium	96	Curium		
45	Rhodium	71	Lutetium	97	Berkelium		
46	Palladium	72	Hafnium	98	Californium		

a few of these familiar elements and learn their symbols. I'll be back in a little while.

Guy looked over the new chart and was excited to see he did know a few elements on the chart. He said, "Here are some elements I recognize—Iron, Copper, Iodine, Mercury, Silver, and Gold." He checked out their symbols on his Periodic Table, using the atomic numbers. Then, he made a list of these elements with their symbols on the back of his Element/Symbol chart exclaiming, "Wow, I am really learning chemistry!

Guy thought, "I should review the elements Atomic Numbers 1 to 20 to make sure I remember them before I start learning new symbols." He put the new chart aside for a while and picked up Professor Terry's Element/Symbol chart. Then he made himself comfortable, and began to review these symbols. He remembered most of them. The ones he forgot, he went over several times.

When he felt he knew them all, Guy decided to learn a few symbols for the elements on the new chart. He learned Mercury, Hg and Iodine, I. Then just as he was having trouble learning the symbols for Silver and Gold, Professor Terry returned.

"Professor Terry, I just keep mixing up the symbols Au, Gold and Ag Silver "

She said, "I have a story that just may help you. I was vacationing in a country where vendors were selling gold chains on the street corners. This one particular vendor kept yelling 'Hey You, want to buy some gold?' The way the man pronounced the words 'Hey you,' sounded like Au, the symbol for Gold. From that time on I've never forgotten that the symbol for Gold is Au. That little man's voice plays over and over in my head. '*Au*, want to buy some gold?' Maybe this story will help you too. I remember Silver is Ag because I associate it with the country **Arg**entina noted for its silver mines."

****

When Guy got back to his family's cabin, Guy knew there were two symbols he would never forget–Au for Gold and Ag for Silver. He looked over the list again and decided to learn at least two more symbols. Guy said, "I think I'll learn Iron which is Fe, and Copper which is Cu."

That night Guy's dreams were about his fantastical trip to Periodic Table Land — the adorable houses, the blinking street lights, the little elements running around. Then all the symbols he had learned swirled around his head until he fell into a deep sleep. He would be ready in the morning for his next adventure in Periodic Table Land.

PART 2
Chapter 2
Guy's Atomic Number Adventure

Guy woke up early. As he dressed he thought of Professor Terry's magical Periodic Table. Sliding down that tunnel of mirrors was so exciting. Best was seeing Periodic Table Land. with its adorable houses the festive colored lights draped around the street sign and those adorable little elements he saw running around. Professor Terry said he would get to meet these little elements. He couldn't wait. The next day took too long to come, but it finally did. Guy dashed down the mountain to the university lab.

Professor Terry said, "Are you ready for a great adventure today? I arranged for you to visit with Sodium in Periodic Table Land. He lives in one of those unique houses that you saw yesterday. Sodium will be your teacher today. I think you will learn a great deal from him."

"Absolutely, I'm more than ready!" Guy could not contain his excitement.

"Well, let's not waste a moment." Professor Terry led Guy dramatically to the back of the lab to the Periodic Table. They both leaned against it, prompting the wall to spin around. Bells, whistles and sirens screamed excitement ahead!

They slid down the tunnel of mirrors reflecting a thousand sparkling lights. Magical sprinkle dust enveloped them like a blanket, and whimsical music played somewhere in the distance. Then, in an instant they found themselves in the fantastical world of Periodic Table Land.

Professor Terry turned toward Guy and said, "Now we need to get to Sodium's house. It's number 11, 1st Street."

Guy remembered that's the way it is here. The streets are what were the Group Numbers and the address on the house is the element's Atomic Number. Sodium is Atomic Number 11 in Group 1A on the Periodic Table. The light green blinking lights on the 1st Street sign beckoned them to Sodium's lovely home surrounded by a colorful garden. It was just beyond Lithium's tiny, but distinctive looking house.

Sodium was standing in his front yard waiting to meet Guy.

Professor Terry said, "Sodium, I'd like you to meet Guy. He's so anxious to learn all about the elements. Guy, this is Sodium. He's got a lot to tell you about himself and his family before he takes you around to learn about the Periodic Table."

Sodium Explains the Element's Color, e⁻ Hair and Body Marking

Sodium said warmly, "Guy, let me tell you about myself. I live with my family here on 1st Street, and to distinguish us from elements in other families, we are all green in color. Each family in Periodic Table Land has a special color and a name. Our family is called the **Alkali Metal Family**. Notice the one electron I have on my head as hair. **Everyone in my family is green and has one electron as hair because we are all in Group 1A**. This means **all the atoms in Group 1A have just one electron in our Outside Energy Level**. That's what makes all the elements in my family so much alike."

"That's how it is here in Periodic Table Land. Each family lives on a different street. All the elements in the same family are the same color and have the same number of electrons on their head for hair. The color and the hair electrons show their Group number on the Periodic Table and how many electrons are in the Outside Energy Levels of each member in that family."

Guy said, "I think I understand. If I see an element of a different color, if he has 5 e⁻ as hair, he's in Group 5A, and his atom will have 5 electrons in its Outside Energy Level. All the elements in Group 5A will look like this."

"Now notice the triangle on my chest," continued Sodium. "Each element in my family, and in all chemical families, has a different mark on his chest. That tells everyone his Period Number which is also the number of energy levels in his atom. I have a triangle on my body. A triangle has 3 sides. So, my symbol tells everyone that I'm in Period 3, and I have 3 energy levels in my atom. Each member of my family is in a different Period on the Periodic Table and has a different number of energy levels in his atom. The mark on his body tells you the Period he is in and also how many energy levels he has." Sodium showed Guy the different marks he would see on the elements' bodies and asked Guy if he understood why each symbol represented the number under it.

Guy said, "I have trouble figuring out why that last symbol means seven."

Sodium explained, "The bar in the middle stands for one energy level and the two arrows at the ends of the bar are triangles. Remember, a triangle has 3 sides, so two triangles is 6 energy levels plus one bar equals 7 energy levels, or Period 7." Then Guy understood that one look at the mark on the element's chest would tell him the Period that element was in, and how many energy levels his atom has.

"I've explained all these symbols to you because you will be meeting so many elements as I teach you about the Periodic Table. When I introduce you to other elements in Periodic Table Land, I want you to notice first how many electrons they have as hair. That will tell you what Group they are in on the Periodic Table, and how many electrons they have in their Outside Energy Level. Then notice the symbol on their bodies. That will tell you the Period that element is in, and how many energy levels the element has."

By this time Professor Terry felt that Guy had gotten to know Sodium well enough. So, she said, "Sodium has a great plan for your day. This is the start of your journey to understand the Periodic Table. I'll leave you two now, and wish you success. You are in good hands, Guy. When you are finished, I'll come pick you up. Back in the lab you can tell me all about your day."

Sodium Tells Guy About His Symbol and His Atomic Number

Sodium noticed Guy looking at his door and said, "We may as well begin right here on my doorstep learning more about me. Do you see the **Na** on my door? It is my **Symbol**. **Na is the first two letters of my name in Latin**. Most of the elements in Periodic Table Land use either the first letter or the first two letters of the English version of their name, but some of our names are from a foreign language. One thing you need to remember is our symbols all begin with a capital letter and the second letter is *always* lower case."

Guy observed the attractive oak door that displayed Sodium's symbol, Na, burned into it above the polished brass plaque displaying #11. Then Guy shared with Sodium that he had memorized the symbols for the first twenty elements.

Sodium was pleased. "That will help you a great deal, Guy. Good job!" Sodium continued his little lesson. "The number 11 on my door is important. It is my Atomic Number. N**o other element in all of Periodic Table Land has this number. This is what makes me special, and I'm proud of that number, as well as being proud of my symbol, Na.** They both belong to me alone." He grasped the knob, and opened the door. "Come on in now."

Sodium offered Guy a drink of cold lemonade as he moved two chairs close together, so that they could continue their conversation.

Sodium said, "I'll share with you now how my Atomic Number makes me special. **The Atomic Number tells everyone how many *protons* are in the nucleus of my atom, and how many *electrons* I *should* have in my energy levels.** I'll tell you some thing right up front. The **number of protons in an atom never changes.** It's the protons that determine the Atomic Number. The element's normal atom has a number of electrons equal to the Atomic Number. There are reasons an atom can gain or lose electrons. For now just know the Atomic Number tells how many electrons an atom *should* have." Sodium continued, "Now let me show you what my atom looks like. If

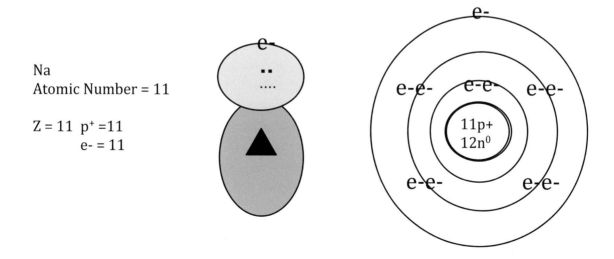

Na
Atomic Number = 11

Z = 11 p$^+$ =11
        e- = 11

you're ready, look at my back wall."

Guy saw the projected image of Sodium's atom on the rear wall of the room. He noticed immediately the 11 protons in the nucleus. Then he counted the electrons, and said "I see you do have 11 electrons in your energy levels. It's just as you said. Your Atomic Number shows everyone how many *protons* and *electrons* are in your atom. But I have a question. What does the 12 n^0 in the nucleus mean?"

Sodium said, "I knew you'd ask that question. It's the neutrons, the other particle in my nucleus which when added to my protons gives you my mass. Pretty soon you will learn lots more about these neutrons and why Sodium atoms have 12 of them."

"One more question," said Guy. "Is it true that not one other element in all of Periodic Table Land has 11 protons and 11 electrons in its atom?"

"That's right," responded Sodium, "If an atom has 11 protons and 11 electrons, it's a Sodium atom. I'm a Sodium atom, and I'm proud of it. Everyone should be proud of who they are. Each person is special. Always remember that. You are special too, Guy. Everyone is special. I'm Number 11. So If you run into an atom that has 11 protons and 11 electrons, you will know it is a Sodium atom. Our Atomic Number is what makes us special. It's what makes us different from the atoms of all other elements."

"Here's the next fact. I want you to remember that we are going to be using the letter Z as a symbol to represent the Atomic Number. You have to remember this because we are all efficient here in Periodic Table Land. So we use letters, numbers, and symbols whenever we can to save time. I can say my Atomic Number is 11, or my Z = 11. That's a lot shorter, and it says the same thing."

"Got it! I'll try hard to remember that the symbol Z is a short way of writing the Atomic Number of an element."

"This is a great beginning for you. Do you have any questions?"

Guy inquired, "Is this true for all the elements in Periodic Table Land? Do all their Atomic Numbers equal the number of protons and electrons in their atoms?"

"Yes," confirmed Sodium, "That's true, but you don't have to take my word for it. We're going to fly around Periodic Table Land and visit four other elements. You can select any four elements yourself. Then we'll visit these elements and see whether their Atomic Number really equals the number of protons and electrons in their atom. Which four elements would you like to visit?"

Guy thought about it for a short time, and then said, "I'd like to see as much of Periodic Table Land as possible." So, looking at the Periodic Table, Guy tried to pick out elements that were located all over Periodic Table Land. With much deliberation, he finally selected four elements that spanned Periodic Table Land from 2nd Street to 8th Street, the last street in the land. "I think I would enjoy visiting Calcium, Aluminum, Oxygen, and Neon."

Sodium said,"I'll take you to visit these elements. We have a special way to get around Periodic Table Land. Each Chemical Family has its own special colored balloons attached to a magic bubble. The balloons match their family color. Our family balloons are green. The sky looks so pretty when we're all out flying at the same time with our colorful balloons. Our magic bubble and balloons are stored at the head of 1ˢᵗ Street near the tarmac. Let's go."

They walked through the wooded area on a narrow dirt path,enjoying the aroma of pine. Guy had some fun kicking the pine cones. They finally came to the shed where the magic bubble and balloons were stored. Sodium and Guy worked together to drag the bubble to the tarmac. Then they inflated the balloons, and attached them to the bubble. Sodium did the final inspection. He made sure that everything was in order for the flight.

They slid into the bubble, sat down, and fastened their safety belts. Once they were both inside Sodium said, "When we twirl the magic wand, the balloons will rise above the house tops."

Guy was really excited to begin the tour of Periodic Table Land. Sodium continued."When we wave our wand, off we go in whatever direction we point. It is the perfect means of transportation for Periodic Table Land. We fly right over the tops of the element houses or as high as we wish. Here we go!"

"Our first stop is nearby on 2ⁿᵈ Street where we will visit Calcium. It's not far. It's just a hop, skip, and a jump away." Since it was such a clear day and they were not in a hurry they took the balloons way up high.

Guy said, "From up here all the houses look so little but the view is great. Looking back, I can see the colorful Period numbers at the edge of Periodic Table Land. The sticks they are on are really short."

When they were almost to 2nd Street, Sodium said, "Here's the wand Guy. You can take it from here. Guy was excited to think he was flying the magic bubble. The green balloons were making their bubble swing from side to side. It was better than the rides in an amusement park."

Soon, they were over 2nd Street. Guy lowered the wand, and the bubble landed softly in front of Calcium's house. They popped out of the bubble, and noticed the amber colored blinking lights surrounding the 2nd Street sign. Sodium told Guy, "Brown is the color of the family that lives on 2nd Street."

Guy said, "How was it you called the blinking lights amber colored? Now, you're saying the family color is brown. I thought the lights were to match the family color."

"Well Guy," said Sodium, "Amber is a shade of brown. It's a yellowish brown color. Brown comes in many different shades. They use amber because it's a prettier shade for lights."

They knocked on Calcium's door which displayed Calcium's symbol, Ca. "This is the first two letters of his name. Did you notice the symbol began with a capital C, and the second letter 'a' was lower case? So many people forget that the second letter must be lower case." Sodium continued, "His address on the door is his Atomic Number."

Guy saw the number 20 on the brass plaque and said, "If what you said is true, Calcium should have 20 protons in his nucleus and 20 electrons in his energy levels. That's a lot of electrons. I can't wait to see Calcium's atom and start learning the meaning of the Atomic Number."

Calcium opened his door and invited Guy and Sodium into his home. Guy greeted

Calcium saying, "It's so good to meet you. I've been told how important you are to people. You make their bones and teeth really strong!"

Sodium happily explained, "I'm teaching Guy all about the Periodic Table, and today he's learning about the Atomic Number. If it's OK with you, we would like to check out the image of your atom."

Calcium was happy to show them his atom. It was projected on his back wall just as Sodium's was. Being a good host, Calcium offered them something to drink recommending a glass of milk. He said, "You know milk is a good source of Calcium which makes your bones strong." Guy and Sodium drank the milk as they checked out Calcium's atom.

Calcium said, "I have a pretty big atom compared with all the atoms you aregoing

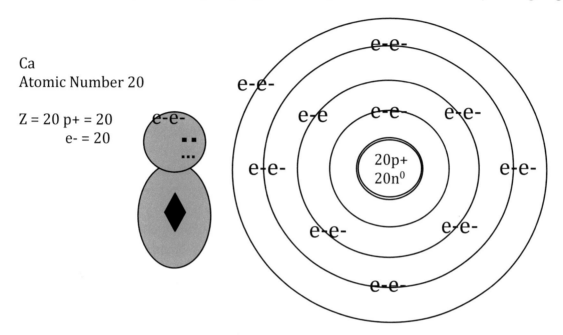

Ca
Atomic Number 20

Z = 20 p+ = 20
e- = 20

to visit. I actually have the most electrons of all the elements you will visit."

Guy saw that Calcium had 20 protons in the nucleus. Then, he counted the number of electrons in the energy levels, which added up to 20. **Calcium's atom had just the same number of electrons and protons as his Atomic Number.** This is what Guy had come to see, and he was happy that they did equal his Atomic Number.

Guy also noted that Calcium had 2 electrons on his head, which meant he was in Group 2A and he counted 2 electrons on the Outside Energy Level of his atom. Looking at the Periodic Table, he saw that Calcium was in Group 2A.

The four sided diamond on Calcium's body told Guy that Calcium was in Period 4. Just to be sure Guy counted the number of energy levels in Calcium's atom. It was 4, Then, he took out his Periodic Table and Calcium was in Period 4.

Guy thanked Calcium for being so kind. Then, he and Sodium walked back to the beautiful bubble with its light green balloons. At the click of their safety belts, Sodium pointed his wand, and up they rose above the housetops. They were on their way to visit the other three elements Aluminum, Oxygen, and Neon.

****

The journey to Aluminum's house was quite long, as they had to traverse the extremely wide B Avenue, the airspace of the Transition Metals. Their metal houses were truly shiny. Professor Terry pointed out the homes of Silver and Gold. They were quite elaborate.

Guy took over the wand and flew the bubble for a while. He looked down and saw Iron, symbol Fe and Copper, symbol Cu running around and playing tag. It was fun to watch.

When they finally reached 3rd Street, where Aluminum resides, the green balloons set the bubble down gently in front of Aluminum's house. The symbol, Al, marked Aluminum's front door just below his shiny brass plate, exhibiting Aluminum's Atomic Number, 13.

Guy said, "Let's see if his protons and electrons are equal to his Atomic Number."

Aluminum greeted them and invited them in. His walls were made of shiny aluminum, and his cozy couch and chairs were upholstered in colorful, plush blue and red plaids, that reflected off the walls. He brought them into the kitchen, where all of his appliances were also made of aluminum. Out of the oven he pulled an aluminum cookie sheet covered with a batch of chocolate chip cookies to share with his guests.

Delighted, Guy accepted the sweet treats, as he told Aluminum how much aluminum his mother had in their family kitchen saying, "We use aluminum pots, pans, cookie sheets, and aluminum foil all the time. It's so nice to see you in person," said Guy, smiling as though he was meeting a familiar friend after a long absence.

"Today Guy is trying to verify that the Atomic Number truly equals both the number of protons in your nucleus as well as the number of electrons in your Energy Levels," explained Sodium.

While Aluminum was busy setting up the projector to put his atom's image on the back wall, Guy noticed that he had 3 electrons for hair. Guy knew he was in Group 3A. At that moment Aluminum succeeded in projecting the image on the wall, and Guy saw the 3 electrons in his Outside Energy Level. The triangle on his body tells me that he has 3 energy levels in his atom, thought Guy. He counted and there were 3.

When Aluminum had everything ready he said, "Take a look at my atom Guy and check out my protons and electrons."

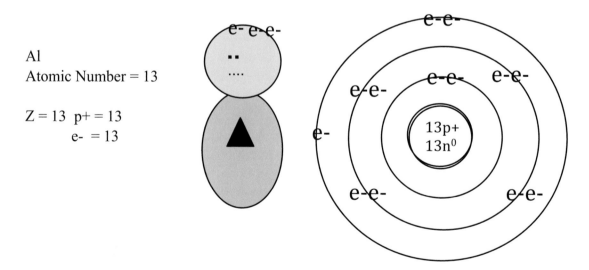

Al
Atomic Number = 13

$Z = 13$ $p+ = 13$
$e- = 13$

Immediately Guy saw the 13 protons in the nucleus of the atom, and counted Aluminum's 13 electrons. This confirmed once again that **the Atomic Number indicated the number of protons and electrons in the atom.**

"Thank you Aluminum, it was good visiting with you."

"And thank you for the cookies," Guy called out as he and Sodium headed for the magic bubble. Once inside the bubble, seat belt fastened, Sodium twirled the wand. The light green balloons rose high in the sky, heading toward 6th Street the home of the element Oxygen. Guy looked down on the uniquely designed element houses nestled among trees and colorful gardens. "What a delightful way to learn about the Atomic Number!" exclaimed Guy.

****

Pointing the magic wand, Sodium directed the balloons to set the bubble down in front of Oxygen's house. The air was extra fresh around Oxygen's home. "I guess," said Sodium, "that's because Oxygen is the gas our lungs need to breathe, and Oxygen has an extra supply of that gas around his house." Sodium added, "Oxygen is Atomic Number 8

and notice his symbol is just one single letter, O."

Guy nodded as Oxygen greeted them and invited them into his house. Inside was also like a breath of extra fresh air. Oxygen asked, "What brings you to visit?"

Sodium spoke up. "Today I'm teaching Guy about the Periodic Table. I want him to be sure that it is universally true that the Atomic Number equals the number of protons and electrons in an element's standard atom. Guy wants to see that your Atomic Number equals the number of your atom's protons and electrons."

"Everyone knows that. That's how they decide what the Atomic number is. They count the protons in the nucleus." Oxygen laughed. "Did you explain that we sometimes call our Atomic Number, our Z number?"

Guy spoke up, "Yes, I notice your Z# is 8."

"Nicely, done! 8 is my Atomic Number," said Oxygen." I guess you do know what Z means." He handed both Sodium and Guy a fan to move that fresh air around, as the day was becoming increasingly warm.

While Oxygen was setting up to project his atom on the back wall, Guy decided to share with Sodium what he had observed as he looked at Oxygen.

Guy started with the fact that Oxygen had 6 electrons on his head for hair. "This meant he lived on 6th Street in Periodic Table Land. which is Group 6A on the Periodic Table." Guy continued, "Best of all, it tells me to look for 6 electrons in the Outside Energy Level of his atom.

"Good job Guy,"said Sodium. "The reverse is true too. If you look at the **Group Number on the Periodic Table, it tells you the number of electrons in the Outside Energy Level of the atoms of all the elements in that Group.**" Next, Guy mentioned that the two dots on Oxygen's body meant Oxygen was in Period 2. "The important thing

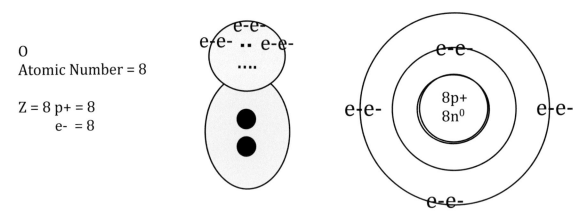

O
Atomic Number = 8

Z = 8 p+ = 8
          e- = 8

to remember is the Period Number equals the number of Energy Levels in the atom." Guy saw that he did have 2 energy levels in his atom.

Oxygen said, "My atom's ready for you to see."

Guy carefully noted eight protons in the nucleus, and then he counted eight electrons in the Energy Levels and finally conceded, "It seems that Sodium is right. **Each element that we have visited has had the same number of protons and electrons in their atom as their Atomic Number.** Thank you Oxygen, for letting us visit and see your atom!"

Guy then reviewed what Professor Terry had taught him about the atom. There were 6 electrons in the Outside Energy Level. So Oxygen was in Group 6A. Indeed the atom had 2 Energy Levels as it should because Oxygen is in Period 2.

****

Sodium said, "Only one more element to visit!" So they slid sideways into the bubble, strapped themselves in, and pointed the magic wand in the direction of 8th Street, Neon's home. The beautiful light green balloons flew them high in the cloudless sky, crossing over many houses. At last, they reached their final destination, the home of Neon.

Neon was out on his porch hanging up a neon sign that he had created to spell out his own name. He was about to hang it near the front door when he heard Sodium and Guy arriving. He had hung other neon signs around decorating his property. Glancing around to make sure his decorative lights met the approval of his artistic eye, Neon spied a sign he had made for Sodium in the shape of Sodium's symbol, Na. It was in the exact green color of the Alkali Metal Family.

From around the corner of the house appeared Sodium with Guy beside him. Neon greeted them."Oh, I'm so happy you have come to see me. I was just looking at something special I made for you. I'll give it to you when you are ready to leave. What's brought you this way today?"

"I'm teaching Guy all about our Atomic Numbers. Could he see your atom?"

"Of course," Neon said, "How exciting!" They went inside Neon's home. It was

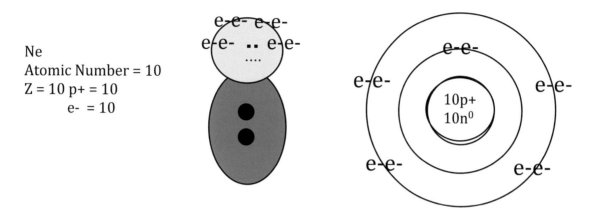

Ne
Atomic Number = 10
Z = 10 p+ = 10
   e- = 10

tastefully decorated with Neon lights. When they reached his living room, Neon switched off the lights and projected an image of his atom on the rear wall.

Guy noticed Neon's Atomic Number was 10 and immediately saw the 10 protons in the nucleus. Then he counted the number of electrons in all the energy levels, and they added up to 10. The protons and electrons did indeed equal Neon's Atomic Number.

Guy said, "Now I'm convinced that **the Atomic Number tells us the number of protons and electrons in an atom.** Thank you, Neon, for finally convincing me that this is true. Neon you should get a prize for being the most colorful home I have seen. I love all the brilliant signs in so many different shapes and sizes that are decorating your property."

Guy then noticed the mark on Neon's body. The two dots meant he had 2 Energy Levels. Guy checked out the Energy Levels in the atom and there were 2. Then he noticed the 8 electrons on Neon's head and checked out the Outside Energy Level of his atom. It did have eight electrons. Guy said, "Neon is in Group 8A, Period 2. I've learned so much today."

They sat around and chatted for a while. Then Guy said, "Now, when I see an atom's Atomic Number, I know that's why the element's atom is different from all others. His atom has exactly the same number of protons and electrons equal to his Atomic Number, and no other atom is like this anywhere." Guy was happy.

As Sodium and Guy prepared to leave, Neon said, "Before you go, I want to give you this neon sign that I made for you." He lifted the green Na sign from the wall and handed it to Sodium. "I made it green to match your Alkali Metal family color." Then, they thanked Neon and were off.

Before returning to Sodium's place, Guy and Sodium zoomed around Period Table Land for a while. With Sodium pointing out different elements that they passed. Guy was able to capture side views and get glimpses of a couple more atoms checking that their Atomic Number equals the number of their protons and electrons.

Sodium said, "If you are going to draw Bohr models some day, you will have to know the number of electrons the atom has in it's energy levels and the number of protons that are in the nucleus. Now you know this."

Guy said, "If I look at the Periodic Table and see the Atomic Number, I'll know how many electrons and protons to put in the atom. Today was truly awesome. I must send a text to each of the elements thanking them for helping me understand the Atomic Number. Thank you, Sodium, for such a super day. The Atomic Numbers on The Periodic Table are so meaningful to me now. It was kind of you to be my teacher."

At that moment, Professor Terry arrived. It was time to return to the lab. They said goodbye to Sodium and slipped back into the university lab. There Guy shared with Professor Terry the story of his adventures flying all around Periodic Table Land and learning about the Atomic Number.

Professor Terry decided that she would test Guy to see how much he had learned. So Professor Terry said, "OK, here's a quiz. You can use my Periodic Table."

Guy ran to the back of the lab to search the Periodic Table for Lithium, Boron and Potassium. Then he wrote the answers to the questions. After Guy was finished, he knew that he had learned what he needed to know. **The number of protons and electrons in an atom is equal to the Atomic Number.**

Atomic Number Quiz

1. What is the Atomic Number for the element Lithium, Li (Group 1A)?____
2. How many protons does the element Boron, B (Group 3A) have? _____
3. How many electrons does the element Potassium, K, (Z=19) have?____
4. What does the symbol "Z" mean?

Answers: 1. Lithium, Z=3. 2. Boron has 5 protons
3. Potassium-19 electrons 4. Z is the symbol for the Atomic Number

Just when it was time to leave, Guy looking quizzically at Professor Terry said, "Now that I learned all about the Atomic Number, I suddenly have lots of questions about those electrons that I've been counting in the Energy Levels."

"Well Guy, you are right to be curious! There is a great deal to learn about the electrons. They are interesting little creatures. So tomorrow, I will arrange to have you study electrons," promised Professor Terry.

That night, Guy dreamed happy dreams about his day's adventures flying around Periodic Table Land. He loved the bubble lifted by the green balloons; the magic wand that pointed it in the right direction, and all those nice elements who let him see their atoms. It was fun. Then his dreams switched to tomorrow. In his dreams the electrons were riding with him in the bubble up high into the fluffy, billowing clouds and back. He had the electrons playing with him around some of those adorable houses. His dreams were just that, dreams. Tune in tomorrow for Guy's real adventures with the electrons.

PART 2
Chapter 3
Guy Meets theElectrons

Professor Terry selected Calcium to teach Guy about electrons because Calcium had so many electrons—20 to be exact. Calcium had agreed, and was getting ready for Guy's arrival. Getting ready is not really a precise description. Frantic preparations is a better portrayal of the scene at number 20, 2ⁿᵈ Street. This is Calcium's address in Periodic Table Land. On the Periodic Table Calcium is located in Group 2A and his Atomic Number is 20. While Calcium was trying to get ready for Guy's arrival to learn about electrons, the electrons were acting up and not cooperating. Calcium's electrons. are always fooling around and having a good time. Today was no exception. Calcium's electrons were running around the house playing hide and seek, laughing, and teasing

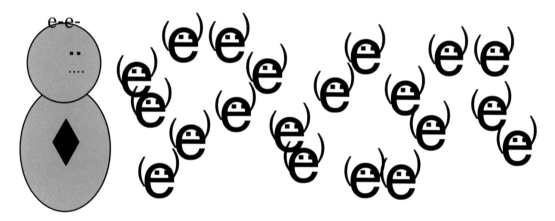

each other. Calcium chided them trying to get his electrons to be serious for a moment, and behave properly.

Meanwhile back on the mountain Guy was getting up. He would soon be heading down the rocky path towards the university. He was curious about what the Professor would have in store for him today. During his whole trek down the mountain he tossed around so many possibilities in his mind. Finally there, he turned the doorknob to the lab and was greeted by Professor Terry's smiling face. He knew she had something really good planned for him.

"Good morning, Guy," said the enthusiastic Professor Terry. "I hope you are ready for Electron Day! I made arrangements with Calcium to teach you. In fact, Calcium wants his electrons to teach you all about themselves."

The concept of electrons actually becoming his teachers was way beyond the possibilities Guy had thought of as he hurried down the mountain path.

Professor Terry put the final touches on the glassware she was organizing on the lab table. Then, with another smile she said, "OK, I'm finished now. Are you ready?"

Guy couldn't contain his excitement about returning to Periodic Table Land. He followed Professor Terry to the back of the lab, and there they leaned against the Periodic Table. The wall spun around and propelled them into the tunnel of mirrors with its thousand sparkling lights. As they slid again toward the fantastical world of Periodic Table Land, they were sprinkled with the now familiar magic dust, and in the distance they heard the mysterious music welcoming their return. In an instant they were standing on a sidewalk in that magical world.

He thought, "How lucky I am to learn chemistry this way!" Off they went to 2nd Street where Calcium lived. As they passed the houses of Beryllium and Magnesium, Guy noted their symbols and Atomic Numbers adorning their oak doors. He was fascinated by the distinctive architecture of each house. Soon they stood in front of number 20, a lovely brownstone home. That was appropriate, thought Guy as Professor Terry had told him that brown was the family color.

Guy said earnestly, "The number on the door is Calcium's Atomic Number. I visited him learning about Atomic Numbers. This means that Calcium has 20 electrons in his energy levels and 20 protons in his nucleus." He wanted to show Professor Terry that he was really remembering the chemistry he had been taught.

Calcium's Electrons Become Guy's Teachers

Calcium came out into the airy way in front of his brownstone house, and leaning on the wrought iron gate said, "I'm so happy to see you both again. Guy, I hear you visited a lot of elements yesterday after you spoke with me. You are now convinced the Atomic Number equals the number of protons and electrons in an atom."

Guy smiled, and nodded in agreement.

Calcium continued, "Professor Terry your message informed me that Guy needs to learn about electrons today." He turned to her and said, "I'll take good care of Guy and make sure he learns everything he needs to know about electrons. You can expect us to be finished about three o'clock."

"Wonderful!" said Professor Terry, and with a friendly wave she left.

Calcium moved closer to Guy and whispered, "My electrons have not been behaving this morning. Wait here while I get them squared away before you come in."

Calcium went inside, and Guy heard him scolding the electrons, "You naughty little electrons stop playing and get back in your proper energy levels. We have company. Guy is here. Act like you know that Guy has come specifically to see you."

The electrons got back in their energy levels where they belonged. See *Diagram A*. Then Guy entered Calcium's house and in chorus, the electrons sang his welcome. "Hi Guy, what would you like to know about us?"

Guy was delighted to hear the little electrons' singing voices. It was a welcome he did not expect. Then Guy responded, "I want to know everything about you! But before we get started, I do have one question that has been troubling me since I read about

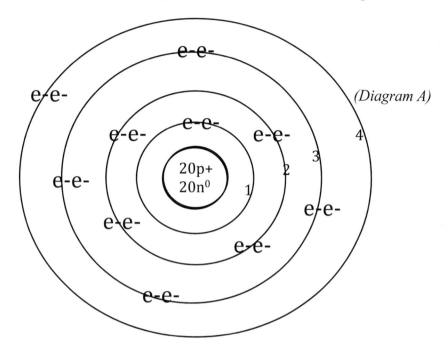

(Diagram A)

electrons in my book. The book said that the electrons were located in an electron cloud around the nucleus. The atomic models I've seen had electrons in special places around the nucleus called energy levels. Which is it?"

The electrons giggled and called out together, "We're in both! We are each in our assigned energy level at special distances from the nucleus. The models you saw were just still pictures of our energy levels. In real life, we are in these specific areas, but we are moving at such high speeds around the nucleus that we look like a cloud. As we move, we remain in areas at our special distances from the nucleus called energy levels. Our appearance is kind of like the way blades of a fan look when they're moving at high speed. The blades appear like just one big blur. Our motion creates the look of a cloud which some people call the electron cloud because we are in there."

"Well now, that makes a lot of sense," said Guy. "Thank you, my little electron friends."

Calcium walked closer to talk to Guy and said, "Electrons are very small, but there are quite a number of things you need to learn about them. My little electrons are excited to be your teachers today."

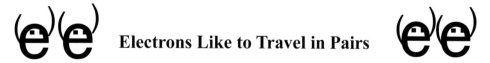

Electrons Like to Travel in Pairs

One little electron piped up, "Did you know we like to travel in pairs? We stay close to our partner as we whirl around the nucleus in our assigned energy level. We like being close to each other. Our negative charges make this difficult, but we have learned spinning in opposite directions overcomes our negative repelling forces. We electrons are clever."

The outside electron objected. "Well not always! If you are in the Outside Energy Level, you might not have a partner. This is true of elements whose Atomic Numbers are not even numbers. Then one electron has to travel alone. I'm lucky. Calcium's Atomic Number is 20, an even number. There are two of us in the outside energy level. We don't have to travel alone. It's so good to have a partner. Look back at the model of our Calcium atom (*Diagram A)* and see all the electrons traveling in pairs. We look like e⁻e⁻ in the energy levels just like we told you."

Calcium whispered to Guy, "These little electrons are amazing."

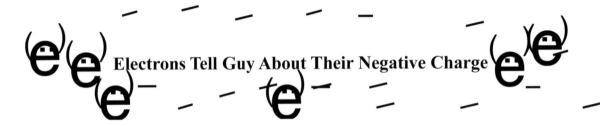

Electrons Tell Guy About Their Negative Charge

The second electron interrupted, "I've got to tell you something exciting about us. Some people are boring, but never accuse electrons of being boring. We are charged! As a matter of fact, we have a negative charge. That's what that little minus sign after the electron (e-) means. Have you ever walked across a rug, touched a doorknob, and suddenly felt a shock? Well, as you walk across a rug, your shoes pick up electrons from the rug and fill your body with extra electrons. What you feel when you touch the door knob is the charge of electrons leaving your body. We electrons are so charged! We're rather fun to be around."

Guy reflected, "So that's why I got such a shock when my cousin shuffled across the rug and touched my nose!" The electrons nodded, and sang, "Kids! Kids! Have so much fun. We electrons give them so much fun!"

Guy Learns About the Protons' Positive Charges

"I hope you tell Guy that we are not the only ones with a charge," said a serious little electron wanting to keep the facts straight. "Protons have a charge, too. That's what the plus sign (+) means when we write protons as p⁺ in the nucleus of the atom. Their charge is a positive charge, but it is all locked up in the nucleus. You don't feel the charge of the proton the way you feel our charge."

"That's exactly my point," the second little electron said, "We electrons just have more fun with our charges."

The little electron continued, "The protons' positive charges are doing something absolutely necessary. I'm sure you know that opposite charges attract each other. It's like the opposite ends of magnets that attract each other. Guy, you probably have played with magnets and experienced the way the magnets pull each other together. We electrons are negative. So the protons' positive charges are attracting our negative charges holding all our electrons in place around the nucleus. For this we should be thankful. It's the positive charges on the protons that keep the electrons with their negative charges from flying off away from the nucleus."

Guy Learns About Periods and Energy Levels

Calcium spoke up, "Guy, the next thing you have to learn is how to figure out how many Energy Levels an atom should have. Who would like to tell Guy about energy levels?"

The outside electrons said, "Whoever's in charge better get the number of energy levels right, or we won't have an energy level in which to race around the nucleus. We definitely need to have the right number of energy levels."

At that, the two electrons on the outer energy level jumped off of their energy level, grabbed Guy's arm and led him through the door and outside saying, "We're going to show you what tells us how many energy levels we need to have." They walked around to the side of Calcium's house. Then they climbed up the stairs to the roof and directed Guy to look far away to the edge of Periodic Table Land. Pointing to the far side of 1st Street which was on the edge of town, the little electrons said, "Do you see all those colorful lights on short poles with numbers on them? Those numbers are the Periods.

PERIODS

1
2
3
4
5
6
7

"There are Period Numbers at the beginning of each row. The Period number tells how many energy levels the elements in that row should have. For example, the number by Calcium's row is 4. This means Calcium should have 4 Energy Levels in his atom."

The electrons wanted Guy to learn this, so they said, "Take a look at your copy of the Periodic Table and find Calcium. Calcium is the second element next to Potassium in row 4."

Guy looked at his Periodic Table, found Calcium and then stood on tip toes and exclaimed, "I see Calcium is really in Period 4. The row starts with Potassium (K) and Calcium (Ca) is next. Then Period 4 goes clear to the other side of the Periodic Table, ending with Krypton (Kr). Wow! All of those elements have four Energy Levels. That's a lot of elements that have the same number of Energy Levels!"

The electrons pointed out to Guy, "You're right. There are a lot of elements in Period 4 because it includes not only the elements in our 8 streets, but also all the elements on B Avenue. I have a picture here of Period 4. You can look at it and see all the elements in Period 4."

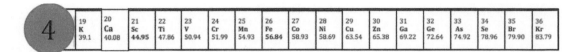

Guy said, "When I noticed that Calcium had a four sided diamond on his chest, I guessed that meant he had 4 energy levels. Now, I see by looking at the Periodic Table that he really is in Period 4. That tells me he has 4 Energy Levels in his atom."

Then, they led Guy back into the house to look at Calcium's model atom projected on the back wall, (*Diagram B*) "Count how many Energy Levels Calcium has."

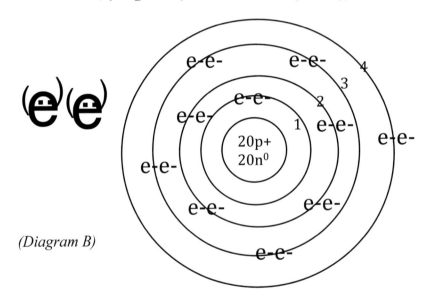

(Diagram B)

Guy carefully counted four Energy Levels, as the little electrons hung the numbers 1, 2, 3, and 4 on the energy levels. "I guess I'll never forget that the Period tells us how many energy levels an element has in his atom."

"Hooray," they squealed with delight. "Now you know why we have four energy levels. We are located in Period 4 on the Periodic Table."

Then Guy said, "Lithium has two energy levels, because Lithium is in Period 2."

Just to make sure that Guy understood what the Periods meant, the little electrons gave Guy one more challenge. "What Period is Aluminum in?"

Guy knew where to find Aluminum on the Periodic Table. That element was right after B Avenue. He observed that when he was flying there to learn about the Atomic Number. He found Aluminum, ran his finger over to the left side of the chart and discovered that Aluminum was in Period 3.

The little electrons squealed with delight again, singing, "Guy's got it! Guy's got it! **The Period tells the number of energy levels an element has.** We did our job teaching him well. Yea Guy! You understand what the Period Number means." They were so happy.

Electrons Are Restricted to Certain Energy Levels

The impatient electrons next to the nucleus couldn't wait another minute. When they saw Guy they said, "Where have you been? We have a lot of important information about the places electrons can be located in the atom. We need you to listen to us. Electrons cannot go where they please. Each energy level can only have a special number of electrons. My partner and I are the only two electrons allowed in this first energy level next to the nucleus." Look at *Diagram C.* "

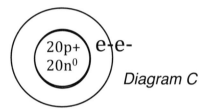

Diagram C

"The nucleus knows we are special. That's why we get an energy level all to ourselves. All the other energy levels have so many electrons in them that they are not as private as we are. We have the whole energy level to ourselves. Energy levels two and three can have up to 8 electrons in them. Energy levels after those might have to squeeze in 18 to 36 electrons."

One rather vocal electron with eight electrons in his energy level spoke up. "You may be special, but we are never lonely. When we have lots of electrons in our energy level, we always have someone to play with."

"OK," Calcium said, "I need you all to behave. You are all important to me. Each of you is special. Remember, I need all 20 of you electrons since my Atomic Number is 20. So, each one of you is important to me. Each of you was made to do something special and special you are. So none of this bragging. Guy knows you are not suppose to brag. Be good!"

Guy said, "My parents have told me that everyone is special. So I know each electron is special no matter how many electrons are in their energy level."

Calcium said, "While we're teaching Guy about Energy Levels, I'm going to show Guy what each of my energy levels is like. I'll do it one energy level at a time. So listen up. It begins with my Atomic Number telling me the total number of electrons I can have.

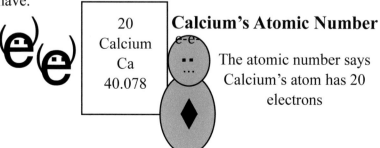

| 20 |
| Calcium |
| Ca |
| 40.078 |

Calcium's Atomic Number

The atomic number says Calcium's atom has 20 electrons

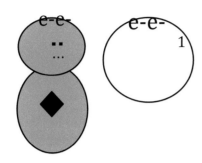

Calcium's Energy Levels

Energy Level 1 -

"This is my first energy level. It is located right next to the nucleus which is in the center of my atom. There are only 2 electrons in this small energy level. This energy level is complete when it has 2 electron in it."

Energy Level 2

"My second Energy Level has 8 electrons. Add this to the 2 in energy level 1 and I now have 10 electrons. Remember I need 20 electrons."

$$2 + 8 = 10$$

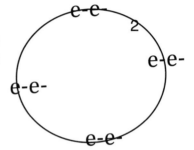

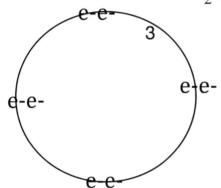

Energy Level 3

"There are eight electrons in Energy Level 3. Add 8 to the 10 electrons we had after Energy Level 2 and now there are 18 electrons in my atom."

$$10 + 8 = 18$$

Energy Level 4

"I have 2 electrons in my Energy Level 4. I needed 2 more electrons to make 20."

$$18 + 2 = 20$$

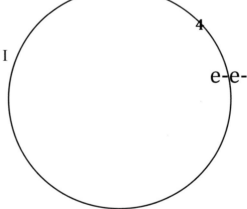

"My 4th and last Energy Level could have eight electrons, but an element can only have a total number of electrons equal to its Atomic Number. So my 4th Energy Level has only two electrons. That's all the electrons needed to give my Calcium atom the 20 electrons my Atomic Number says my atom should have."

"Now, let's look at the model of my atom and see where my 20 electrons are located when it's all put together as a diagram of an atom." They walked over to the back wall where the model of Calcium's atom was projected. *(Diagram D)*

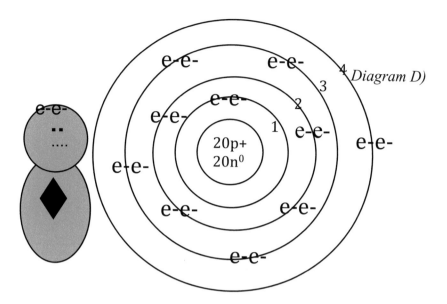

Diagram D)

Excited, the electrons in the Calcium's Outside Energy Level said, "Look! We have two electrons in our Energy Level. We must be as special as those electrons in the first Energy Level."

"No," said the electrons in the first Energy Level, "You have only two electrons because Calcium didn't need any more electrons. If Calcium needed more electrons, you would have had to have them in your energy level. In the first Energy Level, all other electrons are excluded. We are the special ones, the only ones allowed in here."

"Enough!" chided Calcium. "I thought we ended this discussion, letting you know all little electrons are special in one way or another. Actually, I'm about to tell you how you electrons in the Outside Energy Level are really special, no matter how many electrons you have in your energy level."

The Groups and The Outside Energy Level Electrons

"I have a Periodic Table that is all about you electrons in the Outside Energy Levels. You are the reason we have Group Numbers. It's the Outside Energy Level that tells everyone the element's Group Number! That's right, it's **the elements in the Outside Energy Level who determine the Group Number!** You are important. When you see my Periodic Table you will know what I'm talking about. Let's look at my Periodic Table." *(Diagram E)*

Calcium continued, "The Group Number is found at the top of the column in which you find an element. That **Group Number tells how many electrons are in the Outside Energy Levels of each of the elements in that column.** Look at Group 1A on the Periodic Table. *(Diagram E)* All the elements from H (Hydrogen) down to Fr (Francium) have one electron in their Outside Energy Level. That's what the Group number means. There are colored circles below my Periodic Table showing the Group Number and the number of electrons in their Outside Energy Levels. Look at the green

(Diagram E)

circle for Group IA . It has 1 electron on it because all the elements in Group 1A have 1 electron in their Outside Energy Levels."

"Look at any colored circle at the bottom of this Periodic Table. Notice the Group Number tells the number of electrons in the Outside Energy Level. If you look at the gray circle, Group 4A, it has 4 electrons. Group 8A has 8 electrons on its golden circle. Check them all, Guy."

Guy checked them all and found that **the Group number equaled the number of electrons in the element's Outside Energy Level**.

Then Guy said, "Calcium, I remember that your atom had 2 electrons in the Outside Energy Level. *(Diagram D)* That means Calcium should be in Group 2A." Guy checked the Periodic Table and found Group 2A. He searched the column under Group 2A, and there was Calcium. "I guess it's true that the number of electrons in the Outside Energy Level will tell us the element's Group Number."

Then Calcium challenged Guy, "Look at this atom in Diagram F and tell me its Group Number."

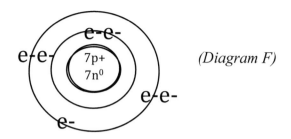

(Diagram F)

Guy said, "I see that this atom has 5 electrons in the Outside Energy Level. That means this element will be in Group 5A."

Calcium said, "This is what I want you to learn. **If you know the number of electrons in the Outside Energy Level of an atom, you will know what Group that element is in on the Periodic Table.** The opposite is also true. **If you know the Group number it tells you the number of electrons in that element's Outside Energy Level.**"

The Atom's Clues About the Element's Name

Guy said, "That atom contains clues to help me discover which element it is."

"**First**, I know the element is in Group 5A because it has 5 electrons in it's Outside Energy Level. So I can look for the Group 5A at the top of one of the columns on the Periodic Table. I know the element is one of the elements in that column."

"**Second** I can count the number of Energy Levels and know what Period it's in. This atom has 2 energy levels. So it's in Period 2. I can now run my finger down to Period 2 in Group 5A and that's the box for this element.

Third Just to be sure I found the right element, I can check if this element's Atomic Number matches the number of protons in the nucleus of the atom. The Atomic Number is 7. That means the atom should have 7 protons. If it does I will know I found the right element in Group 5A, Period 2. It did match. So I found the right element.

Fourth I'll then look at my Element/Symbol Chart and find the name of the element whose Atomic Number is 7. The element's name is Nitrogen."

Guy said,"Boy, look how much I've learned! I feel like a detective. I put the clues together and I solved the mystery of the atom's name. I guess it's true that the number of electrons in the Outside Energy Level will tell us the element's Group Number. That and the Period number will get me the name of the element."

This is how Guy learned what the Group Number had to do with *electrons*. He was convinced that **the Group Number tells us how many electrons are in the Outside Energy Level of an atom**.

"Please remember this fact about the elements in the Outside Energy Levels, Guy," urged Calcium. "It's important because there's even more. Later you will find out what these electrons in the Outside Energy Levels have to do with chemical families, forming compounds and chemical reactions."

At that moment, Guy realized that the electrons in the Outside Energy Level had many important functions. They were special.

Irene P. Reisinger

Displaying the Electron Locations in an Atom

Guy gathered his belongings as he prepared to leave. Suddenly, all the electrons in one chorus sang out, "Don't go, Guy. You can't go until you learn a short way to display our Energy Levels without drawing a whole atom."

A pair of electrons in the middle Energy Level said, "How about giving us a chance to teach. We haven't had a chance to say anything at all."

Guy said, "I'd love to hear what you have to teach me. I heard everyone in Periodic Table Land likes to do things in the shortest way possible. I do, too. So it will be nice to learn a short way to display Energy Levels. If there is a shorter way of displaying Energy Levels than drawing the whole atom, I would love to learn how to do that. I'm not in a hurry to leave. I'll sit here and try to understand what you have to teach me."

The first little electron drew the whole atom of Calcium (*Diagram G*) marking the Energy Levels 1, 2, 3, and 4, starting at the energy level next to the nucleus. Then he put Calcium's 20 electrons in these Energy Levels, placing 2 in the first Energy Level, 8 in the second Energy Level, 8 in the third Energy Level and finally, 2 in the Outside Energy Level. Guy saw it was taking too long. He finally finished. He said, "Now you can see where Calcium's electrons are placed in the atom. "Look at the atom I drew. It tells you so much." (See *diagram G*). The little electron was very proud of the diagram he drew.

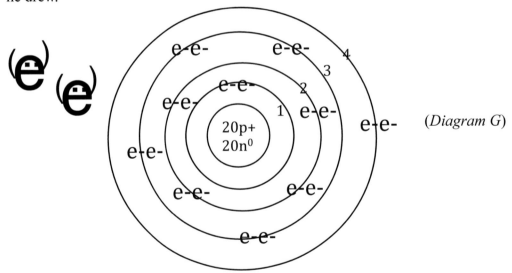

(*Diagram G*)

The second little electron said, "Did you see how long it took my partner to draw all those circles? Then he had to put the 20 electrons in their proper energy levels. Now, I'll show you the shorter way to show the number of energy levels we have, and how many electrons are in each energy level. Watch me!"

"I just grab a pencil and draw a little curved line to represent each energy level that the atom has. Then I write the number of electrons that should be in each energy level in front of the curve. "

84

"Here's how to do it for Calcium, step by step." continued the little electron. "First, I find Calcium on the Periodic Table using his Atomic Number. Then, I look to see what Period Calcium is in. Calcium is in Period 4. That means he has 4 energy levels. So I draw four curved lines to represent the four energy levels."

The little electron wrote—Ca))))

"That's what I call fast,"said the electron with a sense of accomplishment in his voice.

"You're right!" said Guy. "Drawing a curved line to represent each of the energy levels is much faster than drawing the whole atom."

The little electron continued, "Then you put Calcium's 20 electrons in their proper energy level. **You have to remember how many electrons are allowed in each energy level: two in the first and up to eight in the others. You stop when you reach the element's Atomic Number.** So here's Calcium's Energy Levels." Look how I drew it."

Ca 2) 8) 8) 2)

"Now add up the number of electrons, and notice it does equal 20, Calcium's Atomic Number."

Guy said, "This is certainly easier. It does give me a picture of the number of electrons in each energy level, and it's much faster than drawing the entire atom. Thank you for showing me how to do this."

Calcium said "Here's how to display electrons for other elements. I've made a chart for you showing electron models for 4 different elements. It was easy to do. Find the Period and Atomic Number on the Periodic Table. Remember the Period equals the number of Energy Levels, and the Atomic Number equals the number of electrons in the atom.Then remember how many electrons go in each level.

Element/ Symbol	Period	Energy Levels	Atomic Number(Z#)	Electron Model
Hydrogen, H	1	1	1	H 1)
Neon, Ne	2	2	10	Ne 2) 8)
Sodium, Na	3	3	11	Na 2) 8) 1
Potassium, K	4	4	`19	K 2) 8) 8) 1)

Just then two enthusiastic electrons came into the room with another chart they made. "Here Guy, we made you this chart so you won't forget anything we taught you."

Guy gave them a hug and thanked them for the chart and all that they had taught him.

Before Guy had a chance to sit down and read the chart Professor Terry returned to pick him up. It was 3 o'clock, and she was exactly on time. Guy commented on the fact that Professor Terry was always on time.

She took the opportunity to teach Guy a lesson, "Always remember to be on time, Guy. People expect you to be on time. So always make sure you do everything you can to be on time. It takes planning to make this happen."

Calcium reported to Professor Terry, "Guy has learned the most important facts about electrons. My little electrons did a great job."

After thanking Calcium for all Guy had learned, Calcium's 20 little electrons ran up to Guy and gave him a hug, saying, "Guy we had a great time teaching you. I hope you will come back and visit sometime. Don't forget to read the chart we made you."

They left and all the way back to the lab, Guy shared with the Professor what he had learned. Back at the lab, Guy showed Professor Terry the short method he had learned to show how electrons are placed in the energy levels. He used Sulfur as an example. Guy said, "Sulfur is Atomic Number 16. So he has 16 electrons, total. I looked on the Periodic Table and found Sulfur in Period 3, He has 3 Energy Levels. So I drew 3 curved lines. Then I put Sulfur's 16 electrons in their proper energy levels like this—

S 2) 8) 6).

We don't have to draw the whole atom. It's so fast this way, and it provides the same information about Sulfur's number of electrons and where they are in his atom."

Guy was bursting with all the knowledge he had learned about electrons. "They

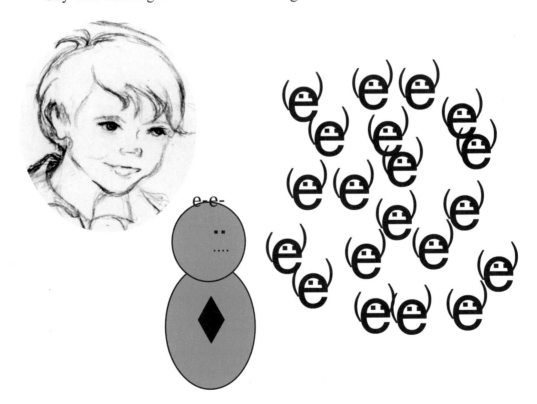

taught me that Periods tell us the number of energy levels in an atom. Then I learned that the Group Number indicates the number of electrons in the Outside Energy Level of an

atom. I learned that electrons move so fast they look like a cloud. I never knew that before. Here's the chart that the electrons gave me. This will tell you all I learned."
'

 ELECTRON FACT CHART

1. Electrons *like* to travel in pairs. That's why they are written e-e-. However, in the atom of an element with an 'odd number,' Atomic Number, one electron has to travel without a partner in the Outside Energy Level, (e-). The minus sign means electrons have a negative charge.

2. **The number of protons in the nucleus of an atom determines the Atomic Number.** The Atomic Number's symbol is Z. It also determins the maximum number of electrons an element can have in its atom.

3. **The number of electrons in each energy level is limited.** Only two electrons are allowed in the energy level next to the nucleus. In each energy level after that, there can be only eight electrons. Energy levels are added until the atom contains the number of electrons equal to the element's Atomic Number. Calcium has 20 electrons. He needs 4 energy levels: Ca 2) 8) 8) 2). This rule applies only to elements 1 to 20. You'll learn about elements later that can have up to 18 or 36 electrons in the last couple of energy levels. Some Periodic Tables will tell you how many electrons there are in those large energy levels.

4. The *Group Number* indicates how many electrons are in the Outside Energy Level of all the elements in the column under the Group Number. Calcium is in Group 2A, so he has two electrons in his Outside Energy Level (the Energy Level farthest from the nucleus).

5. All the elements in the same Group are considered members of the same family, because they have the same number of electrons in their Outside Energy Level.

6. Energy Levels are where the electrons are located in the atom. *The Period Number,* found on the left side of the Periodic Table, tells how many energy levels the elements in that row have in their atoms.

Guy and Professor Terry sat down and enjoyed looking at the chart of all the facts Guy had learned from the electrons. Then Professor Terry said, "Lets see what you know about the element Magnesium. I made up a quiz for you. You can use my Periodic Table in the back of the lab. I'll put the answers here on my desk.

Answers: to Electron Quiz

(1) 12   (2) 12   (3) negative   (4) 3   (5) 3   (6) Mg 2) 8) 2)

(7) 2 A   (8) 2

Guy took the Electron Quiz , and went to the back of the lab near the Periodic Table. He climbed up on a high stool, and leaning on the lab table completed the quiz.

Electron Quiz

1. What is Magnesium's atomic number? _____
2. How many electrons does Magnesium have? _____
3. Is the charge on the electron positive or negative?_____
4. What Period is Magnesium in?_____
5. How many energy levels does Magnesium have?_____
6. Show the electron configuration for Magnesium.Mg ___)___)__)
7. What Group is Magnesium in ?_____
8. How many electrons are in Magnesium's Outside Energy Level? _____

Professor Terry came back, and glanced over Guy's answers and said, "I'll tell you right now that you got them all correct. Good job! However you should get that pad with the answers on it. and check for yourself. It's on my desk."

Guy said it was easy.

1. Magnesium's (Mg) is in Group 2A. So it had 2 electrons in its Outside Energy Level.
2. The Atomic Number was 12, telling me the total number of electrons was 12
3. I looked over to the left side of the Periodic Table and saw Mg was in Period 3 meaning it had 3 energy levels. So I drew 3 curves for Mg)))
4. I put Magnesium's 12 electrons in these 3 curves.
5. Here's the location of Magnesium's electrons.Mg 2) 8) 2)

Guy went over to the desk and found the answers sheet. He read over his answers and saw that he had gotten them all right. Guy said, "That's because those electrons were such good teachers. I learned a lot about electrons, and the electrons taught me to find so much information on the Periodic Table."

Professor Terry was impressed by all Guy had learned.

Guy thanked Professor Terry for sending him to Periodic Table Land to learn about the electrons. Then Guy made the journey up the mountain to his family's cabin.

That night Guy walked down the path to his favorite spot on the mountain and scanned the starry night sky. As the stars and constellations twinkled they seemed to be sending a message to him. "Go Guy! You are discovering the secrets we have hidden away out here in the night sky." Guy felt a peace wash over him as he rested against his favorite rock, and absorbed the awesomeness of the universe. Its immensity and beauty overwhelmed him. He felt more deeply than ever that he had to learn more of the secrets hidden deep beneath the surface of his world. Chemistry is awesome!

PART 2
Chapter 4
Guy Learns About the Atomic Mass Unit

Visiting Professor Terry at the university had become a daily ritual for Guy. On this particular morning when he arrived he was a bit out of breath because he had accidentally overslept. He had raced down the mountain path to the university for his meeting with the professor.

"Have a seat at my desk, Guy," invited Professor Terry when she saw Guy come into the lab flushed and breathing heavily. "I'm just completing an experiment at the moment. While I'm finishing up, you can look at those magazines on my desk."

Guy took a seat at her desk where he could catch his breath looking at the magazines. As he flipped through the pages, he saw diagrams of atoms with their electrons and protons. It was everything he had been learning. He sat there enjoying the captions below each of the pictures. It took quite a while for Professor Terry to complete her work.

Finally, observing that Professor Terry had finished her experiment, Guy asked, "What do you have planned for me today?"

Professor Terry said, "Remember I told you that in order to draw Bohr models you needed to know how to use the Periodic Table to figure out the number of all the particles in the atom. So far, you know the number of electrons and protons in the atom is equal to the Atomic Number. Remember the Atomic Number is sometimes called the Z#. You also know where electrons are placed in the energy levels. Just to make sure you really know this, I have some questions to ask you. First find Potassium's box from the Periodic Table. It's in that magazine you're holding."

Guy turned over the magazine pages, and finally found Potassium's box from the Periodic Table on page ten.

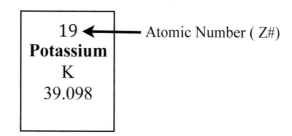

Then Professor Terry asked several questions. "Tell me Potassium's Atomic Number. How many protons and electrons does Potassium's atom have? Show me the location of Potassium's electrons in their energy levels."

Looking at Potassium's box from the Periodic Table, Guy said, "Potassium's **Atomic Number is 19.** Since the number of protons and electrons in an atom equals the

Atomic Number, Potassium has **19 electrons in his energy levels**, and **19 protons in the nucleus** of his atom. Here's where the electrons are located in the energy levels:

K   2)   8)   8)   1)

That adds up to 19, which is Potassium's Atomic Number." Just to show how much more he had learned, Guy added, "I had to draw 4 curved lines to fit all of the electrons. So Potassium ha**s 4 energy levels.** That means it is in **Period 4.** on the Periodic Table. Notice the last curve I drew for Potassium had **1 electron** in it. That's Potassium's Outside Energy Level. So Potassium is located in **Group 1A** on the Periodic Table. I guess I did learn a lot." Guy went to the Periodic Table and found Group 1A. Then he ran his finger down column 1A to Period 4 and there was Potassium. This made Guy very happy. He could use the chemistry he had learned. He knew how to use the Periodic Table.

Professor Terry said, "Today we are going to learn about that other particle in the nucleus besides the proton. What is it Guy?"

Guy was thinking. To help him she quickly drew on the board a diagram of an atom showing the particles in the atom that she had drawn for him the first day they met. Turning to Guy she said, "Does this help?" Guy's face lit up recognizing the answer.

Professor Terry continued, "I see you remembered. The **neutron** is the other

Particles in the Atom(*Diagram B*)

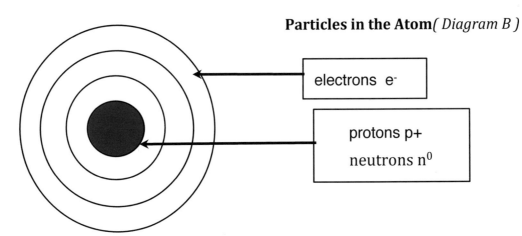

particle in the nucleus that we need to learn about. You already know about the electrons and protons. They equal the Atomic Number. The next step is to learn about the neutrons. They are in the nucleus with the protons. You know how many protons are in the nucleus. Now you need to know how many neutrons each element has in the nucleus of its atom." She drew the nucleus of the Potassium's atom with a question mark indicating we need to know how many neutrons a Potassium atom has in its nucleus.

19 p+
? n^0

The *amu*

"I've got good news and bad news for you, Guy. The bad news is that there is nothing in the element box on the Periodic Table that tells us *exactly* how many neutrons an atom has. However, the good news is that each element box has a number that can help you *figure out* how many neutrons an atom has. It's the *amu*."

"**The *amu*, the atomic mass unit, is that long decimal number at the bottom of each box on the Periodic Table**. Look at Potassium's box on the Periodic Table. Its *amu* is 39.098. When you understand what the amu means, then you will be on your way to being able to calculate the number of neutrons in an atom. Sodium will be your teacher today, and you will learn about the *amu*. For that to happen we must head out to Period Table Land. Let's go, Guy."

Professor Terry and Guy made their way around the slate top lab tables to the back of the lab. Today the air seemed to be saturated with the odor of chemicals, and Guy was thrilled to be in a lab where real chemical experiments were performed. When they reached the Periodic Table, they both leaned on it, prompting the wall to flip around sending them sliding down that mirrored tunnel. The slide to the fantasy world of Periodic Table Land seemed faster than usual. As the magic dust cleared, Guy saw the twinkling colored lights that bedazzled the street signs. Just as Guy remembered, the residents of Periodic Table Land still had the most interesting houses he had ever seen. Each one had a different design. There were mansions, ranch houses, and even a sand castle with a design that was far more beautiful than any sand castle Guy had ever seen at the beach. He was so pleased to be back in this enchanted world. Each time he left, he felt sad at the thought that he'd never get back here again. But here he was again!

Professor Terry said, "Guy, we've come here often enough. I bet you can easily lead the way to the home of our dear friend Sodium."

Guy took the lead, and as Sodium's house came into view Professor Terry said, "I'm running late, so I've got to get back to the lab. Since you are old friends with Sodium, I'll leave you here for your lesson. When you're finished come back to the lab. Are you OK with that?"

Guy went on and arrived at the front gate just as Sodium opened the door to welcome him. Guy turned around, waved goodbye to Professor Terry. She waved back.

Guy Learns About the Isotopes and the *amu*

As Guy shook hands with Sodium, he began to hear sounds he didn't remember from his last visit to Sodium's home. Inside, Sodium's roommates were chattering loudly about something.

"I thought you lived alone," Guy said as he peered curiously around Sodium and into the house.

"No, there are many atoms that live here with me, but we are all Sodium atoms. That means we all have the same Atomic Number. The Atomic Number for Sodium is 11. That means we all have 11 protons (p+) and 11 electrons (e-) in our atoms."

Sodium continued, "Some of us however are a little different. You see, some of us have a different mass, meaning that some of us are a little heavier than others. We call **an atom of the same element with a different mass, an Isotope.** Most elements have isotopes, so it's important to learn about them. Your lesson today will help you learn everything you need to know about mass, which will include understanding isotopes."

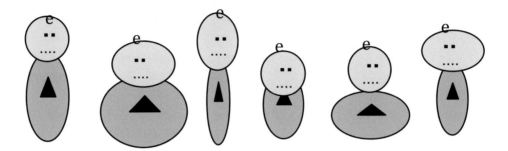

Guy peeked in and saw the Isotopes of all different shapes and sizes.

"Before we go inside, let's start our lesson out here. Take a look at the long decimal number on my doorstep," said Sodium gesturing toward his front step. "That's the average mass of all the Sodium atoms that live with me. We need to get an average, because, as I said before, each of the different isotopes has a mass different from the basic Sodium atom's mass. Chemists measure the mass of all the different atoms, both normal atoms and the isotopes. Then they compute the average mass, which is that long decimal number on my doorstep. You probably never noticed it before, because it's kind of hard to find. It's in a place down low, that is not noticeable. We are required to post it, but we don't like to advertise our mass. Some of us are sensitive about our mass. So, we hope that the long decimal number will not be too noticeable."

"Oh, here it is! I found it—really way down low," said Guy. "I see it now, 22.9698."

From inside Guy heard the atoms call out, "Sodium is telling someone about our mass! Sodium, why do you do that? You know we want to keep our mass a secret. Some of us have too much matter in our body, and we are very sensitive about our mass."

Talking to the isotopes Sodium said, "Don't worry. That number I showed Guy is only the average mass of all of us atoms. It's not any one atom's mass, so calm down. I wouldn't think of hurting your feelings."

Then Sodium went on with his instruction. "When we go inside you will see that the long decimal number that you found on my doorstep is also found at the bottom of my element box on the Periodic Table. It is called the atomic mass unit (*amu*). It could also be called the *a.a.m.*, the average atomic mass, which is exactly what it is: the average mass of all the different kinds of Sodium atoms, our standard atom and our Isotopes."

Guy said, "This must be the number Professor Terry was telling me about. She said that it was going to help me find the mass of the atom that I need to know to calculate neutrons."

"Let's go inside now, Guy." When they got inside, Guy noticed the Isotopes in the back room. They were peeking out like they were sizing him up. They weren't sure that he would understand their different masses. They had heard Guy was a kind person but they needed to be sure.

Guy was looking around to find Sodium's element box from the Periodic Table. He checked out each wall.

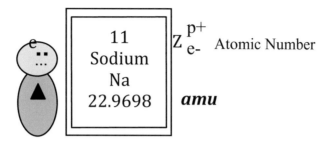

When Sodium finally saw Guy searching, he quickly said, "It's over here on this wall."

Guy looked, and there it was—Sodium's element box from the Periodic Table, framed and hanging on the side wall. The number at the bottom of the box was indeed the number he found out on the doorstep. "This is the number that will help me find that last particle in the atom that I need to know to draw Bohr models—the neutron (n^0)."

"That's right, Guy, I see you found that decimal number at the bottom of the element box. I guess you noticed, it's called the *amu*. The *amu*, as I said before, is the average mass of not only my ordinary Sodium atoms but it also includes the mass of all my Sodium isotopes put together.

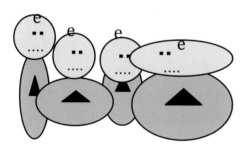

Four of the bravest Isotopes slipped out of the back room and told Guy, "That's right Guy some scientist came around, weighed each of us to get our mass. Then he added all our masses together along with ordinary Sodium's mass and divided by how many of us he weighed. That gave him our average mass which is written at the bottom of Sodium's box on the Periodic Table. It's our average mass so they call it the *amu*, the atomic mass unit. Because it is actually our average mass. It is sometimes referred to as the a.a.m., the average atomic mass of the atom."

When Guy was finished talking to the nice Isotopes, Sodium stepped in to teach Guy something important.

Sodium added, "To find the number of neutrons in an atom you need to know the the atomic mass number, called the A#. Once you know the *amu*, getting the A# of the atom is easy. **All you do is round off that long decimal number, the *amu*, and you have the A#.** The A# is equal to the atom's protons plus the neutrons. Finding the A# is the first step in finding out how many neutrons are in the nucleus of the atom. This is the particle you want to find Guy. One small step at a time. The secret is to learn how to round the *amu*."

Guy Learns How to Find the Atom's Mass Number—The A#

One not so little Isotope came over and said, "I'd really like to tell Guy about the rules for rounding numbers, but first I want to apologize for all of us Isotopes making such a fuss about our mass when you first arrived. You are so nice." All the Isotopes came over and shook Guy's hand. They told him how really sorry they were. "We made up a chart to help you round numbers. I'd like to read it to you." And so he did saying first, "Think of Sodium's *amu* while I read how to round it. It's 22.<u>9</u>97."

RULES FOR ROUNDING

Step #1. Look at the whole number before the decimal point. When you round that number, you will be leaving that number **as it is,** or you will **make it one number larger.** Sodium's *amu* is <u>22</u>.9897. So when it is rounded, the A# will be either 22 or 23. Read about how to decide.

Step #2. Look at the first number after the decimal point—*only* the very first number. Sodium's first number after the decimal is 9—22.**<u>9</u>**897.

Step #3. If the number after the decimal point is 4 or less. Drop the decimal and leave the whole number as it is unchanged. For example, in the number 17.<u>3</u>9999, the number right after the decimal is 3, so drop the decimal. Rounded, the whole number is 17.

Step #4. If the number after the decimal is 5 or more, increase the whole number by one. For Sodium's *amu* 22.**<u>9</u>**697 the first number after the decimal is 9, which is more than 5. Rounded, that would increase the 22 by one. So the rounded whole number for Sodium is 23.

Step #5. The atomic mass number is called the A#. You can write that the mass number of Sodium is 23 or just A = 23. Both are the same.

Guy was so impressed with the isotope's kindness. He said, "Thank you, that was nice of you to go to all this trouble for me. I appreciate it. Your chart will help me remember the rules I need to follow when I round off the *amu* to get the A #"

Then, Sodium wanted to make sure Guy knew how to round numbers, because he would need to round the *amu* to find the mass of the atom when he calculated the number of neutrons in the atom. "Here's a challenge for you, Guy. Can you find the A# of the following elements using the five Rules for Rounding. I've written the answers on this piece of paper, and I'm taping it on the wall. Don't look until you are finished. Then you can see if you are correct."

THE ATOMIC MASS (A) QUIZ

Find the A number for each of the following elements.

Hydrogen *amu* is 1.05 The decimal is .0, which is .4 or less, so A = _____

Aluminum *amu* is 26.91 The decimal is .9, which is .5 or more, so A = _____

Chlorine *amu* is 35.49 The decimal is .4, which is .4 or less, so A = _____

Copper *amu* is 63.53 The decimal is .5, which is .5 or more, so A = _____

ANSWERS: Hydrogen A = 1, Aluminum A= 27, Chlorine A= 35, Copper A= 64. If you got any wrong, ask yourself this question: Did you look <u>only</u> at the first number after the decimal point to decide how to round the whole number before the decimal?

"Great work, Guy! You got them all correct," Sodium said. "Now you know enough that Professor Terry will be able to teach you how to calculate neutrons."

Guy thanked Sodium for teaching him about the isotopes, the *amu*, and how to round the *amu* to get the *atomic mass number* (A). Guy returned to the lab. When he arrived, he could not find Professor Terry. He thought that she was probably attending one of those long meetings she has to go to. So Guy made his way back up the mountain to his parents cabin. He couldn't wait for tomorrow's lesson. He hoped to have the opportunity to journey to Periodic Table Land once more. Soon he would be able to use all that he had learned and draw Bohr models. He was so happy! Drawing Bohr models was his ultimate goal Guy was trying to reach by first learning to interpret the Periodic Table. He was almost there. It would mean he really understood the Periodic Table. First he had to learn how to calculate the number of neutrons in an atom. Maybe I'll learn that tomorrow, he thought. He was full of enthusiasm as he was learning so much chemistry.

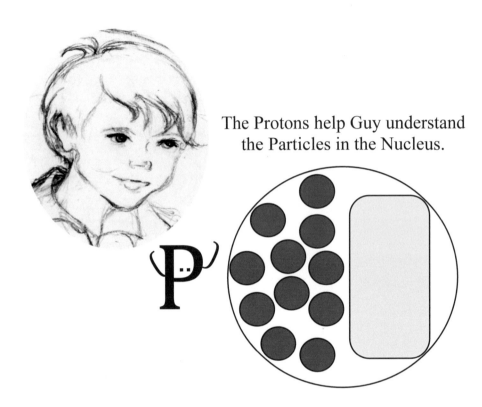

The Protons help Guy understand
the Particles in the Nucleus.

PART 2
Chapter 5
P The Protons Teach Guy About The Neutrons

"Hi Guy!" Professor Terry looked especially pleased to see him.

Guy noticed she had another experiment bubbling away in an Erlenmeyer flask sitting on a tripod with a bunsen burner supplying the heat. The distinctive odor of chemicals, once again, made Guy feel like he was part of the world of science.

"Do you know that yesterday was a milestone on your journey toward understanding chemistry? You accomplished two things. When you learned about the *amu*, you completed your study of the Periodic Table. Then in rounding off the *amu* to find the atomic mass number (A), you were on your way to learning how to calculate the number of neutrons in an atom. That means you have almost reached your goal, being able to use the Periodic Table to find the number of all the particles in an atom. When you have accomplished this you will be ready to reach your ultimate goal being able to draw Bohr models of the elements. To draw Bohr models you need to be able to use the information on the Periodic Table to figure out the number of particles in an atom. That's one of the reason you studied the Periodic Table in such detail. Actually you will need to be able to interpret the Periodic Table as you continue to learn chemistry. Guy, let's hear what you have learned about the Periodic Table in addition to understanding the *amu* which you learned about yesterday."

"I know that **the Atomic Number** is the number at the top of each element box. It tells me the number of protons and electrons in an atom. Next, looking at the top of the columns on the Periodic Table, I see **the Group Number.** There are A and B Groups. There are 8, A Groups. Each tells me how many electrons are in the Outside Energy Level of the atoms of all the elements in that Group. Group 2A elements have 2 electrons there. Group 5A elements have 5 electrons in their Outside Energy Level."

"Finally, I can find **the Period Numbers** down the left side of the Periodic Table. They tell me how many energy levels are in the atoms of all the elements in that row. Period 3 elements each have 3 energy levels in their atoms. Period 6 element each have 6 energy levels in their atoms."

Professor Terry, listening to Guy recount what he had learned about the Periodic Table said, "I hope you are aware of how happy I am that you know how to use the Periodic Table. The atoms of the elements on the Periodic Table are the basic building blocks of everything in the entire universe. You came to me because you were so fascinated by the universe. Now you understand so much that the Periodic Table can tell you about the particles in the atoms that make up your universe. You are on the right path to accomplish your goal of understanding the world you love so much. Aren't you excited?"

Guy thanked Professor Terry for pointing out the significance of yesterday's lesson, as he had not realized it was such a turning point in his chemistry adventure.

"It was significant," said Professor Terry. "Using the information from the Periodic Table will not only make you ready to draw Bohr Models, but later you'll be using that same information on the Periodic Table to construct chemical formulas, understand chemical reactions, and balance chemical equations."

"Finding the mass number of the atom (A) that you learned about yesterday is the bit of information you needed to be able to calculate the number of neutrons in the atom. When you start drawing Bohr models, you need to know the number of all the particles in the atom. It's the particles in the atom that distinguish one element from all others. It's the *amu* on the Periodic Table that is the key to figuring out the one particle you don't know about yet, the neutron."

"Yesterday, you learned that the mass number of the atom (A) is simply the amu rounded off. The protons and the neutrons in the nucleus of the atom are what gives the atom its mass. **All we have to do is subtract the number of protons (p^+) from the whole mass of the atom (A), and we have the number of neutrons(n^0).** It's that easy."

Guy said, "Why don't the electrons count in the mass of the atom?"

Professor Terry said, "The electrons are too small, so their mass does not count."

Guy tried to picture the mass of the atom but he was having a hard time.

Professor Terry said, "Think of the nucleus of an atom like a large peach. The electrons would be like little gnats flying around the peach. The gnats don't add much to the mass of the peach. The electrons add even less to the mass of the atom. The nucleus of the atom is where the atom's mass is. The two main particles in the nucleus are the protons and neutrons. If we want to know how many neutrons are in the nucleus all we have to do is subtract the protons from the mass, A, and we will have the number of neutrons left."

Guy said, "It makes sense, but I can't quite picture it."

Professor Terry said, "Don't feel bad. One picture is worth a thousand words. I talked to Sodium's protons, and they agreed to teach you about the neutrons. They can tell you about neutrons better than anyone since they're in the nucleus with the neutrons all the time. Let's do a quick review."

Guy said, "Yesterday, Sodium showed me how to round that long decimal number in his box on the Periodic Table. That long decimal number is the atomic mass unit, the *amu*, which is the average mass of normal Sodium atoms along with the mass of all his Isotopes. When we rounded the *amu*, we found the A number. That's Sodium's mass number."

Professor Terry said, "The other fact you need to remember is that the number of protons equals the Atomic Number, the Z #. That's the number at the top of the element box on the Periodic Table. Once you know the A number and the number of protons, you can find the neutrons using a formula. It's what I've been telling you. Subtract the protons from the total mass of the atom and you'll know the number of neutrons."

"Let's get going Guy. Sodium's protons will be looking for us." A push on the back wall of the lab and the slide down the tunnel of mirrors brought them quickly to the fantasy world where Guy's been learning chemistry. In no time at all they were entering Sodium's home.

It was obvious that the protons were ready to begin. Sodium directed Guy to look at the model of his atom projected on the side wall.

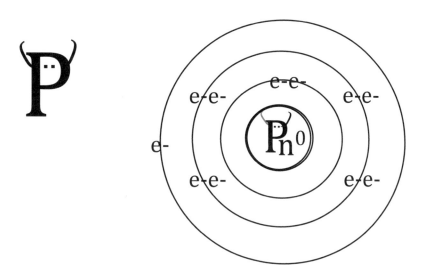

Sodium's 11 protons called out to Guy and said, "Hi Guy! We'll help you learn about the neutrons. Neutrons are our buddies. We live together in the nucleus. If you add us together, you will have the mass of the atom."

"Another way to say it is: if you add together the number of protons and the number of neutrons in the nucleus of the atom you'll find the mass of the atom. Let's set up the card table and get out the play dough and have some fun. Using play dough and construction paper we'll create a model of the nucleus of the Sodium atom," cried the protons as they flew out of the nucleus of Sodium's atom on the wall and landed on the card table.

"Here we are now, ready to make a mock up of the nucleus of the Sodium atom. You're getting it first hand from us protons. We're the experts. So let's get started, but we do need your help".

"Would you please open the red and yellow play dough cans for us? Good! Now, take out **equal amounts of red and yellow play dough because protons and neutrons have about the same mass.** Next give us the lump of red play dough."

As soon as the little protons got it, they started to roll up the red play dough into tiny red balls. Since Sodium's Atomic Number is 11, the protons rolled up 11 little red

balls. One of them said, "You know the number of protons equals the Atomic Number and Sodium's is 11. That's why we're rolling up 11 red balls to represent Sodium's protons."

Guy assured them he did. "I learned that the first day Sodium took me all over Periodic Table Land looking at the atoms of so many elements."

The protons continued their directions. "Now Guy, on that white construction paper draw a circle to represent the nucleus. Make it large enough that these 11 red play dough balls will fit, filling up only half the circle. There has to be room for the neutrons."

Guy drew the circle representing the nucleus on the construction paper.

The protons put their little red balls on one side of the circle that represented the nucleus. Well, almost all the protons did. One naughty little proton decided to play with his ball and have a little fun.

"Hurry now, put your ball in the circle Mister Mischief," said the head proton.

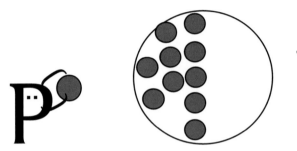

The Number of Protons
equals
The Atomic Number

Reluctantly Mister Mischief did, laughing as he complied. Then all eleven protons were in the nucleus.

"Now Guy, would you please hand us the lump of yellow play dough. Remember we made the red and yellow lumps of play dough the same size because protons and

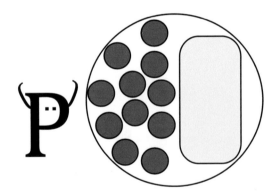

Nucleus of the Sodium Atom
Red Balls = Protons
Yellow Blob = Neutrons

neutrons have about the same mass. This lump of yellow play dough represents the mass of the neutrons. We won't make it into little yellow balls because we don't know how many neutrons there are in the nucleus of Sodium's atom. That's what we are trying to find out. Now put that lump of yellow play dough in the circle right next to the 11 red balls. This is a play dough model of the nucleus of a Sodium atom, Guy."

"Take a look at this nucleus, Guy," said the head proton. "What you see are the red balls and the chunk of yellow play dough. Both of these added together represent the mass of the atom Now think of the red balls and the lump of yellow play dough as protons and neutrons. In real life, it is only the protons and neutrons in the nucleus which give the atom its mass."

"That's great!" said Guy. "Now it's easy for me to imagine the nucleus of Sodium's atom. The 11 red balls are the protons. We knew that there were 11 protons because

Sodium's Atomic Number is 11 and the number of protons equals the Atomic Number. The lump of yellow play dough represents the neutrons. We would like to know how many neutrons there are. Together, the protons and neutrons equal the mass of the atom."

Guy said, "Yesterday, I learned that the mass of the atom is the *amu* rounded off. Sodium's *amu* is 22.9698. Rounding it off, Sodium's mass is 23. So I know that the red balls plus the yellow play dough equals 23, the mass of the Sodium atom."

At that moment Professor Terry walked over to the card table and said, "What would happen if I took away all the red balls?"

Guy responded, "If you removed all the 11 red balls from the circle, only the yellow blob would be left."

Professor Terry continued. "And the yellow blob represents the neutrons. That's the particle we're interested in. We're trying to find out how many neutrons are in Sodium's atom."

Suddenly Professor Terry scooped the 11 red balls out of the circle, saying, "I just removed the protons from Sodium's atom. What's left?"

Guy said, "The yellow blob which represents the neutrons. I think I get what you have done. You took away the protons and what was left was the neutrons."

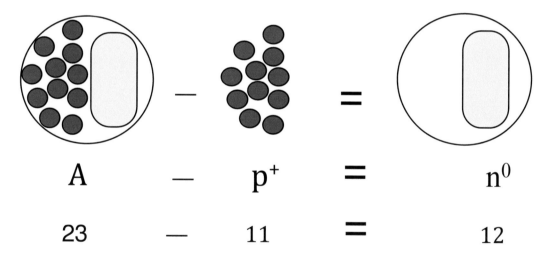

$$A \quad - \quad p^+ \quad = \quad n^0$$

$$23 \quad - \quad 11 \quad = \quad 12$$

"In Math 'take away' is subtraction. So starting with the whole mass which is 23, I take away the 11 protons; now what's left equals the number of neutrons, 12. It's a simple subtraction problem: $23 - 11 = 12$. Twelve is the number of neutrons in the Sodium atom."

Professor Terry commented, "That's all there is to it Guy. You've figured it out. Now you know how to calculate the number of neutrons in any atom. **The number of neutrons (n^0) equals the mass of the atom (A) minus the number of protons (p+) in an atom**. Beneath the diagram above, you saw the formula for calculating the number of neutrons in an atom written like this: $A - p+ = n^0$. Scientists prefer to write the formula this way: $A - Z = n^0$." Professor Terry challenged Guy. "Tell me why they can put Z where the p+ should be in that formula? Hint: Look at Sodium's box from the Periodic Table."

11	
Sodium	
Na	
22.9698	

$Z \, {}^{p+}_{e-}$ Atomic Number

Guy said, "Looking at Sodium's element box from the Periodic Table, I can see that Z and $p+$ are both the same number—the Atomic Number, 11. So both Z and $p+$ are equal to 11. Because the number value of Z and $p+$ are exactly the same, it doesn't matter if we call the protons, Z. The formulas $A - p+ = n^0$ and $A - Z = n^0$ are both the same equations."

"You got it Guy," affirmed Professor Terry. "Scientists prefer $A - Z = n^0$, and we want to be like scientists. So we will use the Z instead of the $p+$."

"One more change that doesn't change anything. Scientists like to put the symbol that you're solving for first. We are solving for neutrons so the neutrons should be first. They just flip the formula around, and it ends up being: $n^0 = A - Z$. This is the formula we will use to calculate the number of neutrons in an atom."

Professor Terry asked Sodium and the head Proton, who had orchestrated the entire lesson, to watch as Guy showed how well he had learned to calculate neutrons. Professor Terry said, "Watch now as Guy finds the number of neutrons in an atom of Chlorine."

**Before Guy Began He Copied
Chlorine's Element Box
From the Periodic Table**

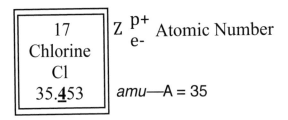

17	
Chlorine	
Cl	
35.453	

$Z \, {}^{p+}_{e-}$ Atomic Number

amu—A = 35

Then Guy said, "Here's the formula that gets you the number of neutrons found in the nucleus of any atom and he wrote $n^0 = A - Z$.

"Next you need to find the number value that Chlorine has for A and Z.

A is the amu rounded off——35

Z is the Atomic Number——17

"**A is 35** because the rule Is : if the decimal is 4 or less the number before the decimal remains unchanged. In 35.453, it was 4, so we kept the whole number 35, unchanged."

"**Z is 17** That was easy. I just copied the Atomic Number from the element box.

Then I did the math for the formula . $A - Z = n^0$ ———$35 - 17 = 18$
There are 18 neutrons in the Chlorine Atom. Guy said, "That was easy."

Sodium and Professor Terry and the head Proton clapped for Guy. "We are all proud of you. You kept saying it was easy. It's easy if you know how—difficult if you don't."

After a short while Professor Terry and Guy left and slipped back into the lab. When they were back there Professor Terry said, "Before you go back to your parents' cabin, I want to show you how to do this problem in a special way."

Guy Learns the Proper Format for Calculating Neutrons

"The Formula Method has four parts." Professor Terry began, "Here they are."

Step 1. Formula: Write the equation using symbols. $n^0 = A - Z$
Step 2. Substitution: Put number values for the symbols. $n^0 = 35 - 17$
Step 3 Calculations   Do the Math $35 - 17 = 18$
Step 4. Answer: Write the answer under the equation. $n^0 = 18$

"This is how you will calculate the number of neutrons for the atoms of each of the elements when you draw their Bohr models. I am not going to require you to show how you did the subtraction. Later, some problems will take a lot of calculations to reach the answer. In that case, I will require you to show all the calculations."

"The neutron was the last particle in the atom that you needed to be able to calculate before you could draw a Bohr model. You are ready to draw Bohr models now. However, I want you to learn something else before I let you begin."

"The lesson, I want you to learn before this happens concerns the meaning of the equation. That will be tomorrow's lesson. You've had enough to think about today. Tomorrow will be your final lesson to prepare you to draw Bohr models of the elements."

Guy was so happy. He knew how to find all the information he needed on the Periodic Table. the Group Number, the Periods, the Atomic Number, the *amu*. He knew the symbols of elements from 1 to 20. He could figure out the number of electrons, protons and neutrons that are in an atom. He was so pleased. He was also tired and went to bed soon after supper.

Guy looked out his window as he was falling asleep and saw the stars twinkling in the sky. They were the same stars that encouraged him to learn chemistry that first night on the mountain. He saw Wish Star waving his magic wand encouraging him to continue his good work. He could hear Wish Star saying, "You can do it Guy! Go Guy!"

His eyes soon closed and he dreamed of Sodium and Calcium and all the other elements he met. He could see their symbols floating around, encircling the adorable little elements that he had come to know and love. In his dreams he was out in space playing with all these elements and their symbols. He was having such a good time. Then he fell into a deep sleep getting ready for tomorrow's lesson.

Guy knows how to calculate neutrons!

$$n^0 = A - Z$$

PART 2
Chapter 6
Guy Learns About the Formula Method

Guy arrived at the lab early, and found Professor Terry sitting at her lab table reading over the observations she had recorded while doing her experiment yesterday. Guy climbed up on one of the tall stools near the Periodic Table and waited for her to look up. While looking at the Periodic Table, he thought about all those elements whose symbols he dreamed about last night. "Soon I'll know what your atoms really look like. Then I will feel that I truly know you."

When Professor Terry finished her work, she walked back to where she observed Guy checking out the Periodic Table. Smiling, she greeted him, "My, you're early today."

Guy said, "Yes, I'm early, I know. I've learned everything I need to know about the Periodic Table to make models of the atoms, and I can't wait to begin. I remember that yesterday you said that I needed one more lesson. You said it was to learn an explanation of the Formula Method for solving science problems. I spent the night dreaming of elements and their symbols. I can't wait to see what their atoms look like."

Professor Terry said, "I want you first to learn the meaning of the formula method. Drawing Bohr models involves calculating the number of neutrons in each of the atoms using the formula method. We used it yesterday, and you seemed to understand how to calculate the number of neutrons in an atom. However, I want you to hear an explanation of the meaning of each of the parts of the equation. Understanding the formula method for solving problems will help you no matter what formula you use."

"I'm ready to hear the formula method explained," said Guy.

"Then let's get started," said Professor Terry and she began by reviewing a simple explanation of the Formula Method. "Scientists **first** write equations using symbols to represent what they need to find out. Then they **substitute** the numbers that they know for the symbols. Once the numbers are in place, they do the **math.** Then they write the **answer** under the equation. Scientific problems can be complex at times. So it's good to learn the formula method for solving problems from the beginning while the math is simple. Later when you run into difficult scientific problems, you will be so used to the formula method that the equation will seem like an old friend there to help you. You will not be intimidated by the difficulty of the problem. It will simply mean that you need to find the right formula, write the equation, substitute the correct numbers in the equation, and solve the math problem to get the answer. It's the same method over and over."

Guy was convinced he needed to hear the parts of the Equation explained if he was going to be like a scientist.

"Once again we are going to visit Sodium—this time to learn about the parts of the Equation. Sodium has another surprise for you too. Let's get started," said Professor Terry as they made their way around the lab tables to the Periodic Table on the back wall. Guy couldn't help being excited each time they did this. Both leaned against the Periodic

Table. The wall flipped around, and they slid down the mirrored tunnel with its thousand sparkling lights. Magic sprinkle dust covered them like a silken blanket and the mysterious music seemed so enchanting. In an instant they slid into the fantasy world of Periodic Table Land. The houses were as adorable as ever. The colorful, blinking lights around the street signs made it so festive. Reaching number 11, 1st Street was a route now familiar to Guy. They were at Sodium's house in no time.

Sodium greeted them cheerfully and said, "Professor Terry, did you tell Guy that you wrote a play to help him learn about the parts of equations?"

"No," replied the Professor, "I thought I'd let you surprise him."

Guy said, "Well, I am totally surprised."

Sodium interrupted and told them that he had all his atoms practicing their parts for this play that Professor Terry had written. "They are back stage now getting ready to begin. I've set up a small theater in my house and invited a whole bunch of electrons, protons, isotopes and atoms that are not in the show to come and enjoy the play with you. So why don't you come in and sit down? You have the best seats in the house, front row center. I'll go back and see if my actors are ready."

Coming from behind the curtain you could hear an interesting mixture of sounds; the strains of violins, the tweeting of the flutes, the tapping of drums and many other instruments in the orchestra tuning up. Mixed in there was an undercurrent of chatter. Finally, Sodium came down the steps on the side of the stage and took his seat beside Guy and Professor Terry. The house lights dimmed, the orchestra played some delightful background music and the play began.

Little Atom #1 emerged from behind the curtain announcing: "Our show's title is

HOW TO SOLVE EQUATIONS THE FORMULA METHOD.

"Welcome Professor Terry, Guy, and our dear teacher and patient director—Sodium. I will be your narrator throughout the play explaining things when I'm needed. Now, let the show begin!" The curtain went up, our narrator moved to stage left.

Little Atom #2
came on stage and moved beside Atom #1 holding the formula poster saying,

$$n^0 = A - Z$$

"This is the formula for calculating the number of neutrons in an atom. My friends, the atoms, will come out holding posters showing the part of this equation that they represent. They will then explain what their symbol means. In the end you will have a better understanding of this equation and you will also have a better idea of what all equations are about. Listen up and enjoy."

Little Atom #3,
holding the poster with the neutron, moved out to center stage and said,

$$n^0$$

"I'm the star of the show. I'm Mr. Neutron. Everyone wants to know how many of my little neutrons are in the nucleus of the atom. You can tell they want to know this because I'm all alone on the left side of the equation. Everything on the right side of this equation tells you what to do to get the answer you are looking for."

"Here's a little bit about me. I'm a neutron. I represent one of the two major particles in the nucleus of an atom. Together with the protons we equal the mass of the atom, A. The zero in my symbol shows I have no charge. You will never get a shock from me the way my electron friends shock you." With that bit of humor, he took his place in the equation line.

Professor Terry whispered to Sodium, "That was cute, but I assure you I did not write that comment in the script." Back stage you can hear a rumble of voices—the electrons objecting.

Little Atom #4
carrying the equal sign stepped to the front of the stage and said,

$$=$$

"I am the equal sign which means everything on the left of my sign is equal to everything on the right of the sign. All you have to do is to perform the math, the equation is asking you to do, and your problem will be solved. In this equations if you do the math the equation is requesting, you will learn the number of neutrons in the atom. That's all I have to say." He then took his place in the equation line.

Little Atom #5

came forward carrying the letter A and said,

A

"I am A, the mass number of the atom. Neutrons make up part of my mass. When you do what the rest of the formula says, you will know how many neutrons are really part of me. "Hey Guy," he yelled out into the audience, "I bet you know what my letter is all about. I saw you learning all about the A number the other day with the Isotopes, under Sodium's direction."

Guy smiled because indeed he knew exactly how to find the A number by rounding the *amu* on the Periodic Table. Guy remembered the short way to do this was to find the *amu* on the Periodic Table. Then 'eye-ball' the number to the right of the decimal and decide whether to leave the whole number the same or raise it up one number. Guy nodded his head in agreement. Changing the *amu*, the average mass of the element and its isotopes, into the mass number of the atom was easy. That's what the symbol A was all about. Guy nodded in agreement. Then Atom #5 took his place in the equation line.

Little Atom #6

came forward holding the minus sign saying,

"My symbol may be small, but it does the work for the whole equation. Without me no one would know what to do with the numbers in the equation. My sign says to subtract what comes after my sign from what is before my sign. I'm a man of few words, but my word is powerful." He took his place in the equation line.

Little Atom #7

came out and moved to the front of the stage holding the card with the Z on it saying,

Z

"My sign Z stands for the Atomic Number of the element. I represents how many protons are in the nucleus of the atom. Remember, the number of protons equals the Z number. If you take me away from the mass of the atom you will have the number of neutrons. I'm the last part of this formula to greet you. Now you know what the whole formula means." Little Atom #7 stepped back into the equation line.

Little Atom #1 the Narrator

came out to center stage and announced, "You Little Atoms did a great job explaining this equation. It was very plain seeing how the parts of the equation work together to solve for the number of neutrons in an atom. Give our little actors a big a hand."

All the Little Atoms 3, 4, 5, 6 and 7 lined up holding their posters in the order of

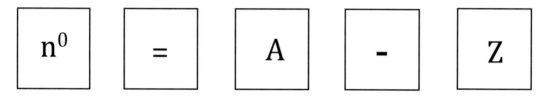

the equation. That way the audience could see once more the equation needed to solve for neutrons. They took several bows and then exited stage right.

Our Narrator, Little Atom #1

returned to center stage saying........

"Wasn't that play wonderful! Everyone clapped again. I have something else to tell you—today is Guy's birthday! What a great birthday for Guy.! He's mastered the Periodic Table. He knows how to calculate neutrons. He's officially ready to draw Bohr models. How perfect that all these accomplishments should come together on his birthday. Congratulations Guy! We are so proud of your accomplishments. Let's all sing Happy Birthday to Guy…...

****

That evening Guy thought back on this wonderful day. This was a perfect birthday. I've accomplished what I have been working for all along. I'll remember today's play whenever I solve for neutrons. When it got dark enough Guy walked up that tree-lined path to his favorite spot, and sat with his head resting back on his favorite rock. He thought back to that first night on the mountain when he looked at the sky and imagined the planets whirling around in space. It was his love for the night sky and the intrigue of the universe that made him want to learn chemistry. He was now well on his way to understanding the secrets hidden deep out there in space. He was happy.

Suddenly, Wish Star jumped down and sat beside him. Wish Star whispered, "You did it Guy. I'm so proud of you. Just know I'm out there watching as you learn more and more chemistry.

Guy said, "I miss you Wish Star. Thank you so much for giving me such a good start. Thank you for giving me Professor Terry and her magic Periodic Table. I've loved my adventures in Periodic Table Land. Thank you for making my wish come true. I'm ready to learn more chemistry—more than ready to draw Bohr models of the elements."

Wish Star said, "I just came to tell you I'm here for you always." With that he spun in a spiral and zoomed off into the night sky.

Guy walked home under a blanket of stars exclaiming out loud with only nature to hear, "What a great summer I'm having! Now it sounds like there's more excitement ahead—Bohr Models, back to Periodic Table Land, and more secrets of the universe to discover. What else could be as much fun as I've been having? I am ready." Back at the cabin, he dreamed of his adventures flying around Periodic Table Land—the elements, those silly electrons, the fun protons, the adorable little atoms. I wonder what those Bohr Models will be like? What adventures lie ahead?

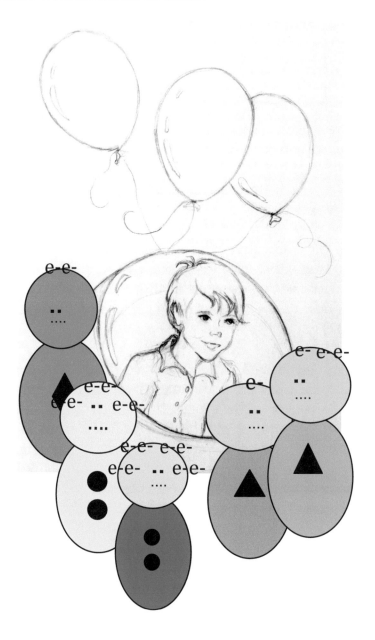

Wish Star says, "You've read **Book 1,** and you know **How To Use The Periodic Table**

Now, it's time to read **Book 2.** Learn to draw **Bohr Models** for the elements so they can see their atoms. Meet the elements who are members of the **Chemical Families**. Discover with Sodium the secret behind **Forming a Compound,** and becoming a Happy Atom. *Continue to learn Basic Chemistry.*

The Happy Atom Story is
MAGICAL !

Printed in the United States
By Bookmasters